Chen Yong
David C. Booth

The Wenchuan Earthquake of 2008

Anatomy of a Disaster

Chen Yong
David C. Booth

The Wenchuan Earthquake of 2008

Anatomy of a Disaster

With 177 figures, 158 of them in color

Authors

Prof. Chen Yong
China Earthquake Administration
63 Fuxing Road
Haidian District
Beijing 100036
China
E-mail: yongchen@seis.ac.cn

Dr. David C. Booth
British Geological Survey
Murchison House
West Mains Road
Edinburgh EH9 3LA
UK
E-mail: dcb@bgs.ac.uk

ISBN 978-7-03-030784-2
Science Press, Beijing

ISBN 978-3-642-21158-4   e-ISBN 978-3-642-21159-1
Springer Heidelberg Dordrecht London New York

Library of Congress Control Number: 2011926988

©Science Press, Beijing and Springer-Verlag Berlin Heidelberg 2011
This work is subject to copyright. All rights are reserved, whether the whole or part of the material is concerned, specifically the rights of translation, reprinting, reuse of illustrations, recitation, broadcasting, reproduction on microfilm or in any other way, and storage in data banks. Duplication of this publication or parts thereof is permitted only under the provisions of the German Copyright Law of September 9, 1965, in its current version, and permission for use must always be obtained from Springer. Violations are liable to prosecution under the German Copyright Law.
The use of general descriptive names, registered names, trademarks, etc. in this publication does not imply, even in the absence of a specific statement, that such names are exempt from the relevant protective laws and regulations and therefore free for general use.

Printed on acid-free paper

Springer is part of Springer Science+Business Media (www.springer.com)

Preface

China is the most populated country on the planet, and also one of the most earthquake-prone. Its history of large and devastating earthquakes, coupled with a long tradition of seismological research, gives it a special place in seismology. Most seismologists would agree that the most significant Chinese earthquakes in recent history are the 1976 Tangshan and 2008 Wenchuan earthquakes. There are many differences between them in terms of location, size, tectonics, casualties, damage, and relief operations. However, the outstanding difference is that, in contrast to the Tangshan disaster, the immensity of the Wenchuan disaster was quickly and widely communicated to the Chinese people and the rest of the world. Fast, open communication of the effects of the earthquake, was implemented for the first time for a major Chinese earthquake disaster. This generated an equally swift and direct response within China and the world, which massively assisted the relief efforts. It has also made possible the speedy publication of this book, giving a unique Chinese perspective on the effects of the earthquake, and the lessons learned.

With a magnitude of $M_S 8.0$, the Wenchuan earthquake is classified as one of the "great earthquakes", which are potentially the most destructive. Unfortunately, since it occurred at shallow depth close to a highly populated area containing many old and poorly constructed buildings, and mountains prone to landslides, its potential for death and destruction was fully realized. To convey the enormity of the earthquake, and show how many lessons are to be learned from the damage to buildings, infrastructure, transport and communication systems, we begin by describing the effects which the powerful seismic waves inflicted on Wenchuan and the surrounding regions. It is probably fortunate that the reader can only get a hint of the thousands of tragedies behind these words and pictures.

Having seen what happened, the next important question is "Why?" Seismologists and engineers must get an understanding of the mechanism of the earthquake, the nature and size of the seismic waves and the ground accelerations they produced, as well as the tectonic movements that generated it, so that the seismic hazard affecting new buildings in Wenchuan and other areas of China can be better assessed and guarded against. China's turbulent tectonic history is certain to continue, with inexorable thrusting and folding of the Earth's crust as India is pushed into Eurasia. We show how the seismic characteristics of the Wenchuan and other earthquakes are monitored, analysed, and used to improve our knowledge of the relevant seismological and tectonic processes.

At the time of the 1976 Tangshan earthquake, through their successful prediction of the 1975 Haicheng earthquake, Chinese seismologists had given the world hope that earthquake prediction might be possible. However, the Tangshan earthquake was not predicted, and neither was the Wenchuan earthquake. Since 1976, decades of research worldwide and the continuing absence of valid predictions of time, place and size (which can not be attributed to chance alone) have demonstrated that the prediction problem is currently insoluble. We describe research on earthquake prediction at Wenchuan, provide a historic perspective on the development of earthquake prediction research in China, and give an indication of its challenges and future direction.

For the foreseeable future, mitigation of seismic risk does not involve prediction, but only the correct assessment of seismic hazard, and appropriate measures for reduction in the vulnerability of buildings and infrastructure. There is no doubt that the effects of the earthquake could have been considerably mitigated if the seismic hazard had been fully recognized, strengthening had been applied to rural buildings as well as urban ones, and public buildings had been given particular attention. You will see a history of seismic hazard research leading the development of better building codes. Sadly, neither had been properly applied in the Wenchuan area. The post-Wenchuan proposals for better risk management as described here, now mandatory and in practice only months after the earthquake, define mitigation practices to be adopted throughout China. When properly implemented, it is these procedures that will save future generations in the earthquake-prone regions of China.

The Wenchuan earthquake caused an unusually high number of deaths from numerous secondary disasters due to its proximity to the mountainous region of Longmenshan. These were caused by landslides, rock falls, mudflows, etc., and their effects on remote villages and transportation systems are discussed throughout the book. Dams, both artificial and landslide-induced, threatened flooding where they to give way. There are many dams in the mountain valleys and some scientists suggested that the Wenchuan earthquake was triggered by the Zipingpu dam and reservoir; this theory is examined in detail.

The exceptionally severe ground shaking and the relatively small proportion of properly strengthened buildings, in a highly populated and mountainous area, caused thousands of buildings to collapse, and posed an enormous challenge to rescuers. The immediate, open and widely broadcast reports and film footage of the scale of the disaster generated an unprecedented response from China and the rest of the world. China's disaster relief acquired many new characteristics, backed by top-level management, such as a people-first open-information approach, and numerous teams of volunteers, offering financial, medical, psychological aids, both from within China and abroad. These new features are described and assessed; they show how the disaster response has made a positive contribution to the health and vitality of Chinese scientific research and social care, and the cohesion and compassion of its people.

The policy of open communication adopted by the Chinese Government when facing natural disaster, has allowed us to describe, assess, and form conclusions on the effects of the Wenchuan earthquake, and to record them here for speedy publication. We have used data, documents and photographs which could not have been released so quickly as in the previous earthquake disasters in China. We hope that geophysicists (particularly seismologists), engineers, social scientists, administrators, and all those who were touched in any way by the devastating Wenchuan earthquake, will find something new and significant in this book.

Chen Yong
David C. Booth
April 2011

Acknowledgements

In 1976, over 240,000 people were killed in the magnitude 7.8 Tangshan earthquake, one of the most devastating earthquakes occurring in mainland China in the 20th century. After the earthquake, Professor Teng-fong Wong, his wife, my colleagues and I wrote a book entitled "The Great Tangshan Earthquake of 1976: An Anatomy of Disaster". The book which was published in English by a British publisher, Pergamon Press, was frequently used by many foreign scholars to understand the earthquakes and earthquake related work in China.

In 2008, a magnitude 8.0 earthquake stroke Sichuan, China, resulting in more than 70,000 causalities and about 1.5 trillion direct economic loss. It was the most destructive, most largely affected, and hardest rescue work involved for an earthquake in China since the 1976 Tangshan earthquake. China and the whole world were shocked by this earthquake disaster. Professor Teng-fong Wong and his wife suggested me to write a book and, as a lesson and experience, objectively tell the most comprehensive information and work experience about this earthquake to the people who care about the earthquakes and earthquake hazard. They also emphasized that tremendous changes had taken place in China during the three decades from the 1976 Tangshan earthquake to the 2008 Wenchuan earthquake. By comparing events in these two large earthquakes, it could show how the Chinese societies have evolved, how the Chinese government and people handled differently these kinds of natural disaster, and how after-earthquake rescue work has been improved. Although Professor Wong and his wife could not join me in the writing of this book because of other commitments, their suggestions were the driving force for writing this book. Here, I express my sincere thanks to them.

This book covers a variety of topics, including seismology, civil engineering, earthquake prediction, management science, humanities and economics, therefore extensive data collection was critical. I must first express my sincere gratitude to the China Earthquake Administration (CEA). This book could not have been written without the support from the CEA's collections of precious and comprehensive data and images, e.g. great Wenchuan earthquake image library, on-the-spot records of great Wenchuan earthquake rescue, and great Wenchuan earthquake scientific survey report. I thank the CEA for the permission of using these data and images. All the data, information and images with no mentioned reference in this book were from the CEA image and data libraries mentioned above.

Many colleagues and friends have helped me. For taking a better picture, they went to the Wenchuan earthquake scene many times. These people include Lu Ming, Guo Xun, Huang Runqiu, Xu Qiang, Liu Ruifeng, Yin Guanghui, Guo Huadong, Gong Jianya. Thanks also go to many more colleagues and friends who I cannot mention here one by one for their help and support. I will remember and thank them from the bottom of my heart.

I also thank my co-author, David C. Booth. He is a senior seismologist in the British Geological Survey, who has long been engaged in the studies of global seismology. He helped polish the English version of this book. More importantly, he used the internationalized language and could objectively appraise the seismological studies in China.

We would like to thank many authors of books and research papers, since a significant proportion of the present book was based on their work. We are also grateful to following people for providing background information and draft assistance, Zhang Wei, Dou Aixia, Chen Qifu, Wang Xiaoqing, Liu Jifu and Shi Peijun, without their help, publication of this book would not be possible.

Chen Yong
April 2011

Contents

1 The Wenchuan earthquake / 001

1.1 Intensity map and damage distribution / 005
Intensity map / 005
Intensity XI and X zones / 011
Strong motion records / 015
General trend of earthquake damage / 017

1.2 Death toll and economic losses / 018
Demography and death toll / 018
GDP losses / 019
Hardest hit area / 021
"Quick and approximate" estimation of casualties / 026

1.3 Earthquake damage to buildings / 028
Damage to buildings according to structural type / 029
Damage to buildings according to age / 033
Damage to buildings according to use / 034
Lessons learned / 036

1.4 Earthquake damage to lifelines / 039
Transportation / 039
Railway systems / 046

Electricity, water supply, sewage and telecommunication / 048

"Quake lakes" / 052

1.5 Secondary disasters: landslides and rock falls / 055

Geomorphology of Wenchuan region / 055

Giant landslides / 062

Mud flows / 064

Rock falls / 066

References / 069

2 Seismological features / 070

2.1 Seismic source parameters for the mainshock / 072

Source parameters / 072

Focal mechanism / 073

Rupture process / 077

Co-seismic deformation / 080

2.2 Historical earthquakes and aftershocks / 084

Historical earthquakes / 084

Aftershocks / 088

2.3 Geophysical investigations before the earthquake / 093

North-South seismic belt / 093

Bouguer gravity anomalies / 095

Crustal thickness / 096

Low-velocity zones / 102

2.4 Tectonic setting of Wenchuan earthquake / 104

Tibetan Plateau / 104

Longmenshan / 107

An intracontinental thrust fault / 111

Weak crustal layer / 113

2.5 Seismic waves generated by earthquake / 115

Seismic waves circle the Earth / 115

Ground motion felt at Beijing / 116

Comparison with waves generated by Tangshan earthquake of 1976 / 118

References / 120

3 Prediction efforts prior to the Wenchuan earthquake / 122

3.1 Earthquake monitoring system in China / 124

National earthquake monitoring network / 124

Other geophysical observations / 127

Data processing and analysis procedures / 127

3.2 Prediction efforts prior to the Wenchuan earthquake / 130

Seismic network of 300 seismometers / 132

Big shock, small GPS displacement / 132

Seismicity pattern / 134

Precursor anomalies / 137

3.3 Development of earthquake prediction research in China / 140

Severe earthquake disasters in China / 140

Organized efforts for earthquake prediction (1966) / 143

China Earthquake Administration (CEA) /144

Haicheng earthquake prediction / 146

Earthquake Act / 150

3.4 Future prospects for earthquake prediction / 152

Realistic public expectation of earthquake prediction / 152

Why is the prediction effort so persistent? / 156

From prediction to mitigation / 158

References / 159

4 Seismic hazard and risk assessment / 160

4.1 Seismic hazard assessment / 161

1957—First intensity zoning map of China / 163

1977 intensity zoning map of China / 163

1990 intensity zoning map of China / 164

2001—Latest hazard map of China / 165

4.2 Building code / 167

Building code (TJ11-78)—effective 1979–1990 / 168

Building code (GBJ11-89)—effective 1991–2001 / 170

Building code (GB50011-2001) / 170

Importance of complying with the seismic design code / 170

4.3 Risk management / 172

Quantification of disaster / 172

Fortification standards should be increased / 174

Earthquake-safe Rural Housing Demonstrations / 176

4.4 Did the reservoir impoundment trigger the Wenchuan earthquake? / 180

References / 184

5 Emergency response and rescue / 186

5.1 People-oriented rescue principle / 188

Immediate top level government response / 188

So long as there was hope, they would never give up / 190

Respect for life / 196

5.2 Open and transparent relief information / 199

Timely, open and transparent information / 199

International relief / 201

Rescue efforts / 204

5.3 The breadth and diversity of donors / 206

Donations from countries and regions / 206

Donations from NGOs / 209

The breadth and diversity of donors / 211

5.4 Voluntary contributions / 213

An individual voluntary response / 215

Rewards of voluntary work / 216

Medical volunteers / 217

Importance of psychological assistance / 218

References / 221

6 Reconstruction of Wenchuan / 222

6.1 Outline of reconstruction / 223

Overall planning / 224

Reconstruction objectives / 226

Reconstruction funds / 227

6.2 Counterpart assistance / 229

New mechanism of post-earthquake reconstruction / 230

From counterpart assistance to counterpart cooperation / 234

6.3 Solid schools, sweet hopes / 235

Schools in reconstruction / 235

Hospitals and other public facilities / 238

6.4 New Wenchuan after the earthquake / 241

All be housed / 241

Life goes on / 242

The main lessons of the earthquake / 245

References / 247

Appendices / 248

Appendix 1 Significant events of Wenchuan earthquake / 248

Appendix 2 Wenchuan earthquake sequence catalog / 258

Appendix 3 Law of the People's Republic of China on Protecting Against and Mitigating Earthquake Disasters / 266

Appendix 4 The Overall Planning for Post-Wenchuan Earthquake Restoration and Reconstruction (Extracts) / 267

The Wenchuan earthquake 1

005 / Intensity map and damage distribution
018 / Death toll and economic losses
028 / Earthquake damage to buildings
039 / Earthquake damage to lifelines
055 / Secondary disasters: landslides and rock falls

The Wenchuan earthquake was the most destructive earthquake in China, with the widest zone of influence and the most serious disaster-induced losses, since the founding of the People's Republic in 1949. Measured at $M_S 8.0$ in magnitude, and XI on the Modified Mercalli Intensity (MMI) scale of felt effects, the earthquake also caused severe secondary disasters such as landslides, mud-rock flows, barrier lakes, etc (Figure 1.1).

1. The earthquake caused large loss of life. On 25 August 2008, it was estimated that there were 69,226 people killed, 374,643 injured, and 17,923 missing.
2. Numerous urban and rural buildings were devastated. Some towns, including Beichuan, Yingxiu, and a vast number of villages were razed to the ground.

Figure 1.1 Under the combination of strong ground motion, surface rupture and giant landslides, Beichuan County suffered severe damage: more than 80% of buildings collapsed.

3. This earthquake made disaster zones of cities in many provinces of the western part of China and about 450,000 km² was designated as a disaster area. According to the Ministry of Civil Affairs (25 June 2008), about 23,143,000 housing units were damaged in the earthquake (where on average, 1 house is equivalent to 4 housing units), and as many as 6,525,000 were destroyed.
4. The infrastructure was severely damaged, and systems providing transportation, electricity, telecommunications, water and gas supply, etc., were paralyzed over a large area. In Sichuan Province alone, 5 national roads, 10 provincial roads and 17,000 country roads were severely damaged. The railway between Baoji and Chengdu, a transportation life-line for rescue and relief, was interrupted for over 4 days. There were no landline communication connections because all facilities were damaged for at least one week after the quake.
5. Public service facilities including schools and hospitals were severely damaged, as well as a large number of cultural and natural heritage sites.
6. Industrial development was greatly impeded, with key industries and numerous enterprises devastated.
7. Large acreage of farmland were destroyed.
8. The quake also wreaked havoc on the eco-environment, with large stretches of forests destroyed, wildlife habitats destroyed, and ecological functions degraded.

The unprecedented intensity of its destructive effects can be attributed to several factors. First of all, the ground motions were very strong. The maximum Peak Ground Acceleration (PGA) recorded at the Wolong station, at an epicentral distance of 23 km (no records exist within 23 km), was close to 1 g (Figure 1.2).

Secondly, before 2008, many counties located in the Wenchuan earthquake fault belt were assigned an earthquake zoning of relatively low intensity for China, such as intensity VI or VII (Chinese intensity scale, similar to the MMI scale, ranging up to XII). Therefore, little consideration had been given to earthquake resistance in the building code. Most of the industrial and residential buildings were highly vulnerable to seismic damage.

Finally, the earthquake occurred in a mountainous region. Losses from

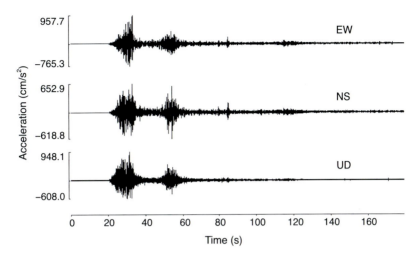

Figure 1.2 Maximum PGA as recorded at Wolong station at 23 km epicentral distance. The PGA was: 957.7 cm/s^2 (EW); 652.9 cm/s^2 (NS); 948.1 cm/s^2 (UD) (after China Earthquake Administration, 2008).

quake-triggered geological disasters (landslides, rock falls, debris flow) accounted for over a third of the total losses, which is extremely rare in the history of earthquake disasters.

We now discuss the earthquake damage distribution according to the seismic intensity distribution, and then consider in more detail the earthquake damage to buildings and lifelines. Finally we discuss the geological disasters, including the earthquake lakes.

1.1 Intensity map and damage distribution

Intensity map

The seismic intensity scale in China is similar to the MMI scale, where intensity is divided into twelve levels. There is an approximate correspondence between the intensity level and maximum ground acceleration as recorded by strong motion instruments (Table 1.1).

An expert group was dispatched to survey 4105 localities over half a million km², where the intensity was evaluated after the earthquake based on the degree of damage of buildings and structures. The intensity distribution map for the Wenchuan earthquake was issued by China Earthquake Administration (CEA) in August 2008 (Figure 1.3).

Table 1.1 Approximate correspondence between earthquake intensity level and PGA (from GB50011-2001, Code for Seismic Design of Buildings)

Intensity	VI	VII	VIII	IX
PGA (horizontal) (cm/s^2)	50	100–150	200–300	> 400

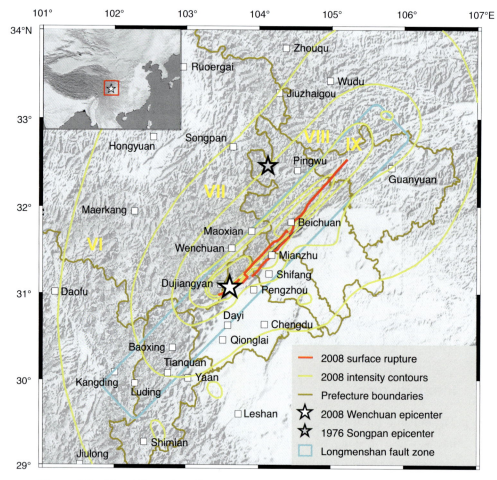

Figure 1.3 Map of the region affected by the 2008 Wenchuan earthquake. Yellow lines are isoseismal lines. Surface ruptures (red lines) due to the Wenchuan earthquake are from Xu et al. (2008). Brown lines outline the administrative areas of Chengdu City and the hardest hit cities and counties (modified from Chen et al., 2010).

The shape of the intensity distribution is that of a narrow ellipsoid, where the ratio of long to short axis may be 10 or greater. The narrow ellipsoid shape of the intensity distribution also describes the damage distribution. Most earthquakes in the eastern part of mainland of China have circular isoseismic contours, such as the Tangshan earthquake of 1976 (Figure 1.4). However, most earthquakes surrounding the Tibetan Plateau have a very narrow ellipsoidal intensity distribution; Figure 1.5 shows the intensity distributions of Tibetan Plateau earthquakes.

From these five cases, it is seen that the narrow ellipsoid shape for isoseismic contours is a common feature of earthquakes on the border of the Tibetan Plateau.

Figure 1.4 Spatial distribution of earthquake intensity due to the Tangshan earthquake of 1976 ($M7.8$). Most earthquakes in eastern mainland China have similar circular isoseismic contours.

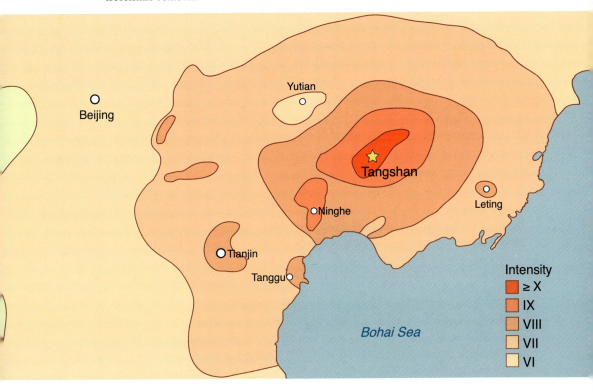

008 | The Wenchuan Earthquake of 2008

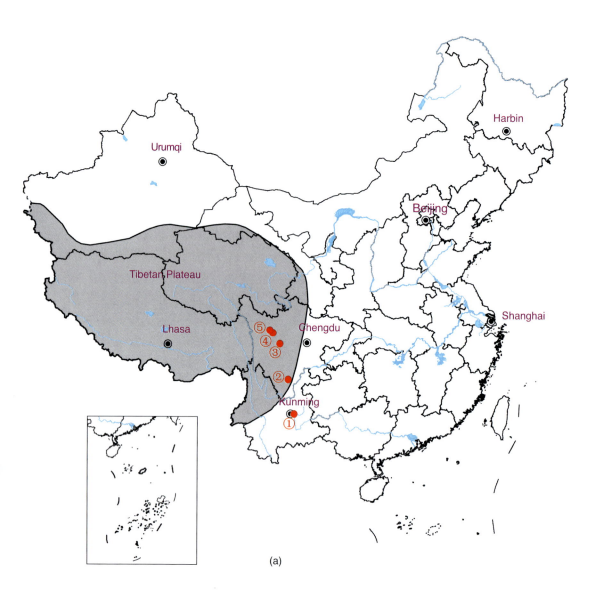

Figure 1.5 Locations and isoseismal contours of five earthquakes around the Tibetan Plateau; all the high intensity isoseismic contours are narrow ellipsoids (intensity colour scales as the same as Figure 1.4):
(a) Circled numbers 1 to 5 represent the locations of five earthquakes in the Tibetan Plateau margin.
(b) Xiaojiang earthquake, 1833-09-06, $M8$, 25.0°N 103.0°E, epicentral intensity XI.
(c) Zemuhe earthquake, 1850-09-12, $M7.5$, 27.7°N 102.4°E, epicentral intensity X+.
(d) Xianshuihe earthquake, 1893-08-29, $M7$, 30.6°N 101.5°E, epicentral intensity IX+.
(e) Xianshuihe earthquake, 1923-03-24. $M7.3$, 31.3°N 100.8°E, epicentral intensity X.
(f) Xianshuihe earthquake, 1973-02-06, $M7.6$, 31.5°N 100.5°E, epicentral intensity X.
(modified from Wen et al., 2008)

Chapter 1 · The Wenchuan earthquake | 009

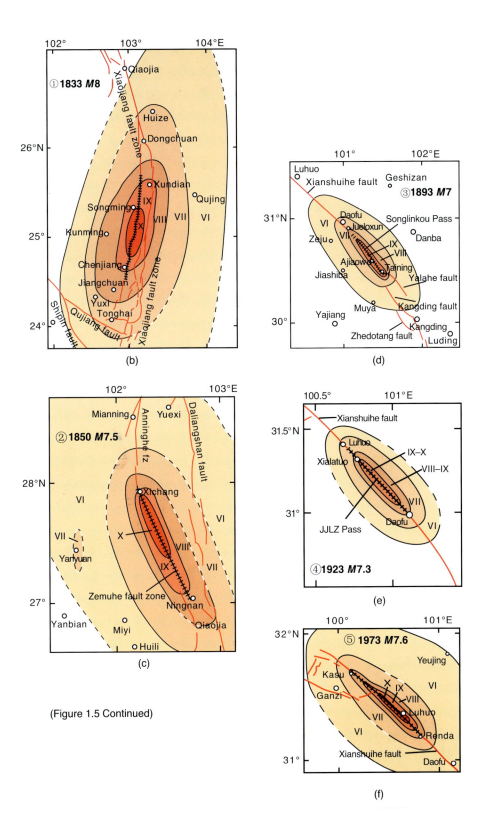

(Figure 1.5 Continued)

Figure 1.6 Damage at Yingxiu which located in the center of XI intensity zone (photo by Chen Kai, Xinhua News Agency).

Intensity XI and X zones

The maximum intensity experienced in the Wenchuan earthquake was XI. The intensity XI zones are two separate ellipses whose major axes follow a NE-trend. The north-east ellipse is centered at Yingxiu in Wenchuan County, and the south-west ellipse is centered at Beichuan. The total area of intensity XI is about 2419 km^2 (Figures 1.6, 1.7, 1.8, 1.9).

Basically all the buildings in this area, including factories, schools, shops, apartment houses and others, were toppled or destroyed (Figures 1.10, 1.11). Tall brick masonry structures (such as chimneys and water towers) either collapsed or broke into several segments. The main structure of bridges collapsed and their foundations were tilted or dislocated. Rail tracks were bent or buckled, and fissures were evident in road surfaces. Ground uplift as well as subsidence was observed.

The intensity X zone was an ellipse with its major axis (224 km) on a NE-trend, with the north-east end of the major axis in Qingchuan County, and the south-west end in Wenchuan County. The total area of intensity X is about 3144 km^2.

The areas of intensity IX, VII and VII are about 7738 km^2, 27,786 km^2, and 84,449 km^2, respectively.

Figure 1.7　Total collapse of brick masonry residences in Yingxiu. (intensity XI)

Figure 1.8 Industrial and residential buildings in the Ming river valley near Wenchuan County. Construction of most village buildings and houses was based on residents' experience instead of controlled, professional design and construction, resulting in severe damage. (intensity XI)

Figure 1.9   Collapsed brick masonry buildings in Beichuan County. (intensity XI)

Figure 1.10   Three-storey buildings with a frame structure in the lower part and brick masonry structure in the upper part, were severely damaged. Hanwang Village of Mianzhu City. (intensity X)

Figure 1.11 Damage to a building of Dongfang Steam Turbine Works (DFSTW) in Mianzhu City. Most of these buildings, built of bent columns, PC trusses, and precast ribbed roof slabs, collapsed entirely, causing serious loss of life and facilities. (intensity X)

Strong motion records

Strong motion data for the $M_S 8.0$ shock were recorded by a network covering Sichuan Province. There were 78 strong motion seismographs located in the Longmenshan fault zone (LFZ), and 49 strong motion recordings were obtained (Figures 1.12, 1.13). Unfortunately, many stations in the epicentral region (particularly at Yingxiu and Beichuan, the intensity XI area) did not work because of instrument failure. The largest PGA up to 957.7 cm/s^2 was recorded at Wolong station at 23 km epicentral distance (Figure 1.2). Six stations were located in the intensity IX area and the PGAs recorded by them are respectively: 641 cm/s^2 (Mianzhu, 87 km); 592 cm/s^2 (Shifang, 64 km); 529 cm/s^2 (Jiangyou, 154 km); 522 cm/s^2 (Jiangyou, Hanzheng, 145 km); 530 cm/s^2 (Guangyuan, 312 km); 430 cm/s^2 (Maoxian, 104 km) (China Earthquake Administration, 2008). These data were used to define and shape the intensity contours.

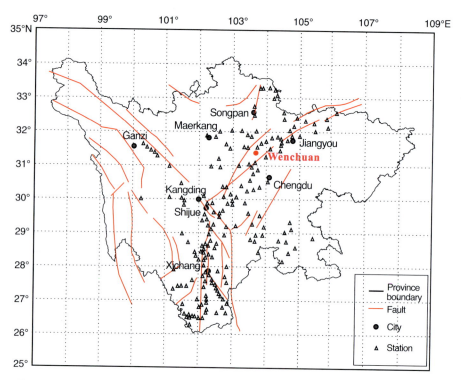

Figure 1.12 Strong motion seismograph distribution in Sichuan Province (after Zhou, 2008).

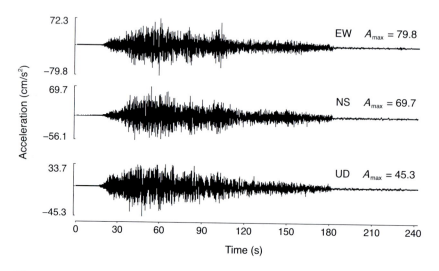

Figure 1.13 Strong motion recorded at Chengdu station located on bedrock. The PGA was between 45.3 cm/s^2 and 79.8 cm/s^2. This corresponds to an intensity value between VI and VII (after China Earthquake Administration, 2008).

The PGA attenuation has very strong directivity: attenuation along the E–W direction is very fast, and along the N–S direction it is very slow (Figure 1.14). The directivity of the strong motion coincides with the directivity of the isoseismic contours.

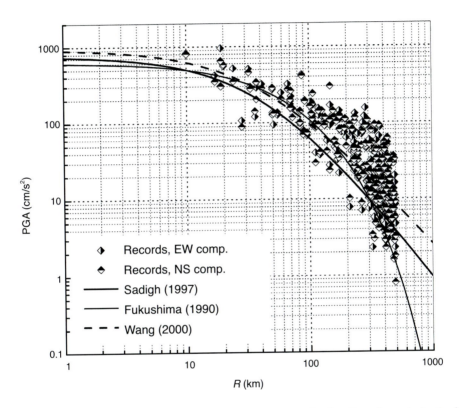

Figure 1.14 Horizontal PGA versus fault distance (R) for data from the mainshock of 12 May 2008 and three attenuation relationships (from Li et al., 2008).

General trend of earthquake damage

It can be seen that there is a general trend to the intensity distribution. The Wenchuan earthquake occurred in Longmenshan fault zone which has a generally north-east trend and is longer than 300 km. Therefore, the intensity contour has strong directivity. For example, the area of intensity IX is like a narrow ellipse with a NE major axis of 300 km, and a SW minor axis of 50 km. The earthquake rupture started in the SW part (Wenchuan County) of the Longmenshan fault system, and propagated NE as far as 300 km (to Beichuan County and Qingchuan County).

In the direction perpendicular to the Longmenshan fault, Chengdu (capital of Sichuan Province) 70 km away from the fault suffered minor damage (intensity VII). But in the direction along the Longmenshan fault, Beichuan 150 km distant from the epicenter suffered an intensity of X–XI. Such strong directivity, influence of the fault, concentration of earthquake damage in a narrow strip are significant characteristics of the Wenchuan earthquake.

Why there is so strong trend of earthquake damage? Most of the earthquake energy was released in a volume (seismic source body) 300 km long, 50 km wide and 20 km deep. Energy release in the source body was extremely uneven, and strong ground motion in the epicentral area shows clear asymmetrical characteristics in time and space. The high intensity areas were confined to the source body. Within this narrow high intensity zone, houses, schools and hospitals etc. suffered severe damage and earthquake triggered landslides. Mud flows and rock falls were also extremely serious.

1.2 Death toll and economic losses

Demography and death toll

According to Chinese state officials, the earthquake caused 69,226 known deaths, of which 68,636 were in Sichuan Province; 17,923 people are listed as missing, and 374,643 were injured. This estimate includes 158 earthquake relief workers who were killed in landslides as they tried to repair roads. The earthquake left at least 5 million people homeless.

The total population of Sichuan Province (Figure 1.15) has been declining, from 84.3 million in 1997 to 81.3 million in 2007. The majority of the province's population are Han Chinese, who are scattered throughout the region with the exception of the far western areas.

The disaster areas are the home of many ethnic minority groups, including China's only concentrated inhabited area for the Qiang ethnic group, and one of the major Tibetan settlement areas, with great cultural

Figure 1.15 Sichuan is a province in Southwest China; its capital is Chengdu. Sichuan means in Chinese "four river courses". Sichuan is one of the biggest provinces of China, with a long history more than two thousands years.

diversity and unique historical and human resources. Significant minorities of Tibetans, Yi, Qiang and Naxi reside in the western portion, forming a traditional transition zone between Central Asian and East Asian cultures. Sichuan has been one of China's most populous provinces.

GDP losses

On 4 September 2008, the Wenchuan Earthquake Expert Committee declared that the Wenchuan earthquake had caused direct economic losses of 845.1 billion yuan (RMB). Property losses involving housing, from large, urban housing developments to rural individual homes, were 27.4% of the total losses. Losses of schools, hospitals and other non-residential premises were 20.4% of the total losses. Infrastructure losses, consisting of roads, bridges and other urban infrastructure, accounted for 21.9% of the total losses. Together, these three components comprised the largest proportion, about 70%, of the total economic losses caused by the Wenchuan earthquake.

Sichuan Province suffered the most serious losses, 91.3% of the total losses, while in Gansu and Shaanxi Provinces, the corresponding figures were 5.8% and 2.9%. Sichuan's economic development level is moderate. Although haunted by economic instability, the degree of industrialization in the plains is comparatively high, while the high plateau areas operate on a much smaller economic scale with few industrial structures and a poverty-stricken population.

Sichuan's nominal GDP in 2007 was 1050 billion yuan (Figure 1.16). The 2008 earthquake losses for Wenchuan (County) are 80% of the nominal GDP for Sichuan Province in 2007. Later in 2008, the Sichuan earthquake further adversely affected the economy of the province. For example, the earthquake reduced the development of the tourism industry, causing a 10.3% decline in revenue. The economy of the province is expected to recover momentum as the government focuses on recovery work and increasing domestic demand. On 6 November 2008, the central government announced that it would spend 1 trillion yuan (about 146.5 billion US dollars) over the following three years to rebuild areas ravaged by the earthquake.

The nominal GDP per capita (pc) for Sichuan Province has risen from 3940 yuan in 1997 to 12,930 yuan in 2007. In 2008, the pc income of urban residents reached 15,630 yuan while rural residents' net pc income was

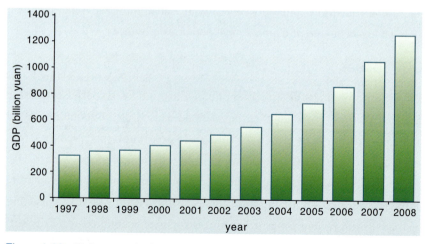

Figure 1.16 Sichuan nominal GDP (billion yuan) as a function of time (from http://www.starmass.com).

4120 yuan. In 2005, urban residents' pc expenditure amounted to 6370 yuan, while rural residents' pc expenditure was 2010 yuan. Two features are apparent. First, the growth in nominal GDP per capita, income and expenditure reflects the rising living standards of the population. Second, although there has been rapid economic growth, large regional differences have remained, in particular between urban regions and rural regions. The Wenchuan earthquake occurred in a large rural region, where buildings and structures had low earthquake resistance. This is the main reason why the Wenchuan earthquake disaster toll was so heavy in human and economic losses.

Many local and international experts found that the massive damage to properties and houses occurred because China had not created an adequate seismic design code until after the devastating Tangshan earthquake of 1976. They established that building codes do now exist, which take care of earthquake issues and seismic design issues. Presumably, many of the Wenchuan buildings were quite old and probably not regulated during construction. As a result, the poorer, rural villages were hardest hit.

Even with many of the largest cities in Sichuan Province suffering only minor damage from the quake, the economic loss runs higher than 125 billion dollars, making the earthquake one of the costliest natural disasters in Chinese history.

Hardest hit area

According to disaster severity (including the death toll, property losses and economic losses), the disaster areas have been divided into three categories of hit, hard hit, and hardest hit (Table 1.2).

Table 1.2 Classification of disaster areas

	Deaths[*] (%)	Houses collapsed (%)	Economic losses (%)
Hardest hit area	97.2	42.9	39.5
Hard hit area	2.0	44.7	44.7
Hit area	0.8	12.1	15.2

* Ratio of deaths in each area to the total deaths.

Hardest hit area 10 counties or cities in total: Wenchuan, Beichuan, Mianzhu, Shifang, Qingchuan, Maoxian, Anxian, Dujiangyan, Pingwu and Pengzhou (Figure 1.17). All of them were located in intensity X or XI regions (Figures 1.18, 1.19). Over 1000 deaths occurred in each of these counties and cities. 97.2% of the total deaths occurred in the 10 hardest hit areas. The total area of the 10 hardest hit areas is about 26,000 km^2, and their locations and names are given in Figure 1.17 and in Table 1.3.

Hard hit area 36 counties or cities, located in intensity VII, VIII, and IX regions in Sichuan Province, Gansu Province and Shaanxi Province. The

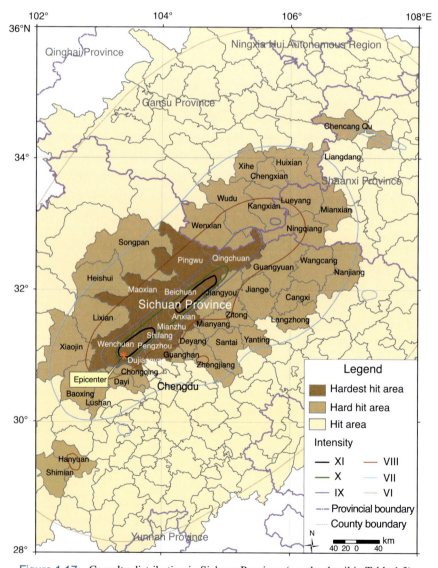

Figure 1.17　Casualty distribution in Sichuan Province (see the detail in Table 1.3).

Figure 1.18 Qushan Village of Beichuan County, located in the hardest hit area.

Table 1.3 Disaster effects in the hardest hit area

County or city	Population (×10^4)	Intensity	Deaths	Collapsed houses
Wenchuan	11	VIII–XI	23,871	608,198
Beichuan	16	IX–XI	20,047	347,856
Mianzhu	51	IX–X	11,380	1,397,925
Shifang	43	VIII–X	6,132	1,006,921
Qingchuan	25	VIII–X	4,819	714,084
Maoxian	11	VIII–X	4,088	300,229
Anxian	50	VIII–X	3,295	774,896
Dujiangyan	61	IX–XI	3,388	655,265
Pingwu	19	VIII–X	6,565	299,557
Pengzhou	78	VIII–X	1,131	622,066

total area is 90,000 km^2, and the economic losses in the hard hit area were about 44.7% of the total.

Hit area located in VI, and VII intensity regions; the total area of the hit area of Wenchuan earthquake is 380,000 km^2, including 186 counties or cities.

Figure 1.19 Beichuan County cemetery, 7000 victims of Wenchuan earthquake were buried here. A huge lawn was marked "5·12" words (image by UAV shot, CEA; courtesy of Lu Ming).

"Quick and approximate" estimation of casualties

A key question which must be addressed in immediate post-earthquake disaster reduction is "how many people may have been killed by the earthquake and what losses may a city or region have suffered?" When the telephone, mobile phone, communications and transportation (roads and railways) are severely damaged, it is very difficult to quickly get information on casualties and damage.

Figure 1.20 and Table 1.4 give the estimated numbers of dead and

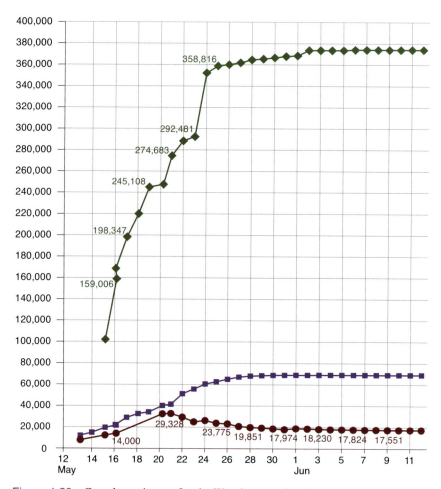

Figure 1.20   Casualty estimates for the Wenchuan earthquake as a function of time. Blue squares: deaths; green squares: injured; brown squares: missing. The number of deaths was given as 14,866 on the second day after the mainshock, and became 69,226 one month later. The number of injured was 100,000 the second day after the mainshock, and was close to 380,000 one month later.

Table 1.4 Reported casualties after 2008 Wenchuan M8.0 earthquake up to 2 June 2008

No.	Report M-D	Time H-M	Time from mainshock	Death toll	No.	Report M-D	Time H-M	Time from mainshock	Death toll
1	5-12	16:00	1.5	4	13	5-20	18:00	195.5	40,075
2	5-12	18:00	3.5	107	14	5-22	10:00	235.5	51,151
3	5-12	21:00	6.5	157	15	5-23	12:00	261.5	55,740
4	5-12	22:00	7.5	592	16	5-24	12:00	285.5	60,560
5	5-12	22:50	8.3	8,533	17	5-25	12:00	309.5	62,664
6	5-13	17:00	26.5	11,921	18	5-26	12:00	333.5	65,080
7	5-14	14:00	47.5	14,866	19	5-27	12:00	357.5	67,183
8	5-15	16:00	73.5	19,500	20	5-28	12:00	381.5	68,109
9	5-16	14:00	95.5	22,060	21	5-29	12:00	405.5	68,516
10	5-17	14:00	119.5	28,891	22	5-30	12:00	429.5	68,858
11	5-18	14:00	143.5	32,477	23	6-01	12:00	477.5	69,016
12	5-19	16:00	169.5	34,073	24	6-02	12:00	501.5	69,019

injured caused by the Wenchuan earthquake as a function of time.

While the earthquake parameters of the Wenchuan earthquake (where, when, and how big) were available a few minutes after the earthquake, it was a few weeks before the number of deaths and the total economic losses were known. This situation has occurred in many other big earthquake disasters, and it is usually hard to estimate the deaths caused by a great earthquake. For example, a similar situation occurred after the Kobe, Japan earthquake (M7.2) of 17 January 1995. Kobe is the second most populated area after Tokyo in Japan, with a population of 10 million. Kobe shook for only 20 seconds, but in that short time, over 5000 died and 100 billion dollars of damage was done (Figure 1.21).

Fast estimation of the casualties after an earthquake would provide valuable information for deploying and managing relief work, and make it more efficient. Liu et al. (2005) proposed a model for estimation of the casualties:

$$N = N_0 [1 - \exp(-\alpha T)]$$

N_0 is the final casualty number, and N is the number at time T, here α is a constant which depends on the earthquake. If we can rapidly obtain α after

The Wenchuan Earthquake of 2008

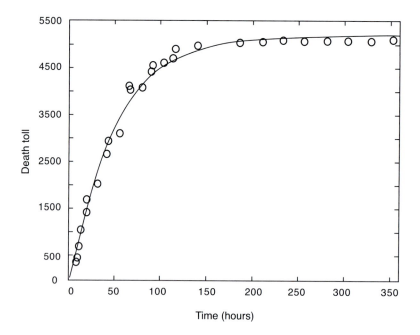

Figure 1.21 Reported deaths due to the Kobe earthquake, Japan, 1995 (circles) plotted against number of hours after the mainshock. Note that the estimate in the first two days was only 1600, one third of the final total (from Wu et al., 2009).

the earthquake, using the number N of deaths reported time T, the final death toll can be estimated in a relatively short time.

Chen et al. (2002) developed a simplified approach for seismic hazard and risk analysis. Using information on the Wenchuan earthquake issued by China Earthquake Administration (CEA) at 12 minutes after earthquake, from the preliminary estimate of magnitude as 7.8, Chen and his colleague gave an estimate of 24,000 dead.

We believe that a new branch of seismology—emergency seismology— could be developed in the future which would be capable of giving a "quick and approximate" estimate of casualty numbers in near real time.

1.3 Earthquake damage to buildings

According to the Ministry of Civil Affairs (25 June 2008), about 23,143,000 housing units were damaged in the earthquake, and as many as 6,525,000

collapsed. Why was the damage caused by the Wenchuan earthquake disaster so severe? The main reasons are:
- Strong ground shaking
- Low earthquake resistance
- Severe secondary disasters (in mountainous areas): landslides etc.

Shaking and ground rupture are the main effects created by earthquakes, and these cause damage to buildings and other rigid structures with different degrees of severity. On a local scale, the severity of local effects depends on a complex combination of earthquake magnitude, distance from the epicenter, and local geological and geomorphological conditions, which may amplify or reduce seismic wave amplitude.

A expert group was dispatched to survey the safety of buildings 10 days after the earthquake (Civil and Structural Groups of Tsinghua University, 2008). The main purpose of survey was to collect the information of houses and buildings damaged by earthquake in disaster areas. Much of the material in this section was drawn from their survey report. The group divided the degree of buildings damage into four levels: (1) minor damage, useable; (2) moderate damage, requires strengthening or repair for further use; (3) serious damage, further use prohibited; (4) dangerous, must be demolished otherwise could topple at any time.

Damage to buildings according to structural type

Buildings in Wenchuan and the surrounding area can be categorized into three main structural types.

1. Brick masonry structures

Most brick masonry structures were two to five storeys high (exceptionally up to eight storeys). For example, as shown in Figure 1.22, in the earthquake-hit areas, most civil buildings were of this type. The form of principal damage can be described as follows: for brick masonry walls, cracks were liable to occur due to lower tensile and shear strength and poor ductility. In cases of slight damage, minor cracks occurred near the corners of door/window openings. In case of more severe destruction, diagonal shear cracks or X-shape cracks occurred on both the long and transverse walls, on the walls between and under windows, or over whole walls. In case of very severe destruction, the walls collapsed or broke apart. For buildings with open or partially open bottom frames, due to the large variation in rigidity (rigid upper part, flexible

Figure 1.22 Beichuan County, located on the central part of Longmenshan fault zone, suffered devastating destruction. This five-storey brick masonry residential building in Beichuan suffered severe damage, did not collapse, and had to be removed.

lower part), the ends of concrete columns were seriously damaged, or the frames even collapsed completely, resulting in subsidence of the whole building. For building with inner frames, due to differences in deformation between the frames and infilled walls, cracks were liable to occur in the walls. Tie-columns and ring beams played an effective and integral role in constraining movement, though some cracks occurred at the connecting ends, where concrete was crushed. High blocks were damaged to varying degrees, or collapsed, due to the whipping effect of the shaking. Staircase walls were liable to be damaged at openings for electricity meters. Walls were partially destroyed and attached signs or other decorative material fell off due to pounding effects of buildings on both sides of expansion joints which were not wide enough.

A selected group of 201 brick masonry buildings were inspected after the earthquake for seismic damage. Of these, 42 suffered minor damage and could be used, 74 suffered moderate damage but required strengthening or repair for further use (Figure 1.23), 33 suffered serious damage and were banned from use. 52 structures were assigned for demolition as they were dangerously unstable (Table 1.5).

Figure 1.23 Most ancient structures in the earthquake affected areas had mortised timber frames, grey tiles, and brick or stone enclosure walls; most ancient pagodas were close-eaves brick pagodas. The main damage to ancient pagodas was cracking of the pagoda body and collapse at the top. The seventeen-storey Kuiguang pagoda in Dujiangyan City, a hexagonal brick tower, is shown. It shows cracking from top to bottom, and must be strengthened or repaired for further use.

Table 1.5 Damage statistics for various structural types (data from Civil and Structural Groups of Tsinghua University et al., 2008; modified)[*]

	Minor damage: can be used	Moderate damage: repair required for further use	Serious damage: further use prohibited	Demolition essential
Brick structure	42(21%)	74(37%)	33(16%)	52(26%)
Mixed	20(48%)	9(21%)	4(10%)	9(21%)
Steel-reinforced concrete frame	66(63%)	40(38%)	8(8%)	9(9%)

* 366 buildings were investigated.

2. Steel-reinforced concrete frame structures

In the quake-hit areas, reinforced concrete structures contained open frames, frame shear walls, and frame infilled walls. In the frame structure, most damage occurred at the column ends and beam-column joints (representing a strong beam and weak column) (Figure 1.24). The

Figure 1.24 Beichuan middle school: five-storey reinforced concrete frame structure. Entire collapse of ground floor, and upper 4 storeys damaged severely. Assigned for demolition.

main forms of damage were: cracks at column ends and joints, buckling of reinforcing bars, and bending cracks or diagonal shear at the beam ends. For shear wall structures, damage included shear failure of the spandrel and damage to the lower part of the shear wall. For frame infill wall structures, the main forms of damage forms were: diagonal or cross-diagonal cracks on the end walls and at the corner of door openings; horizontal cracks above or below window openings, wall collapse in buildings with large rooms, large window openings, and arch walls. Damage to the stairs in frame structures was recognized as a new problem. In previous practice, only static stress analysis was performed on the stairs without considering seismic effects. However, a stair adds lateral stiffness to the frame structure, and the stair slab suffers repeated tension and compression under the horizontal seismic movements.

3. Mixed brick and frame construction

We shall call these, "brick-frame" structures. An example is an eleven-storey insurance company building in Jiangyou City. In this steel-reinforced concrete frame structure, no damage occurred to the frames, but some cracks appeared in the wall surface, so that it required repair before further use.

Damage to buildings according to age

The expert group investigated the damage level as a function of the date of construction. There are three national building codes in China: TJ11-78 (issued in 1978), GBJ11-89 (issued in 1989), and GB50011-2001 (issued in 2001). Design standards of the building codes have increased with time, and with China's economic development. Table 1.6 shows that structures built before 1978 were seriously damaged, because then the economy was relatively poor, most buildings were made in brick, and their earthquake resistance was lower than buildings built after 1978 (Figure 1.25).

Table 1.6 Damage statistics for various construction dates (data from Civil and Structural Groups of Tsinghua University et al., 2008; modified)[*]

	Minor damage: can be used	Moderate damage: repair required for further use	Serious damage: further use prohibited	Demolition essential
Before 1878	5(10%)	19(38%)	4(8%)	21(43%)
1979–1988	21(35%)	20(33%)	8(13%)	11(18%)
1989–2001	35(40%)	27(31%)	14(16%)	12(14%)
After 2002	32(52%)	19(31%)	3(5%)	7(11%)
unknown	44(35%)	45(36%)	16(13%)	21(17%)

* 384 buildings were investigated.

Figure 1.25 Complete collapse of old brick masonry house, Hanwang Village in Mianzhu City. Most village houses were built using experience rather than controlled professional design and construction. As a result, the earthquake caused massive destruction to these houses and enormous losses to life and property.

Damage to buildings according to use

Four types of building were used in the statistical survey: school buildings, government offices, residence houses and industrial buildings (Figures 1.26, 1.27). The total number of samples was 484 (Figure 1.28). Schools and industrial buildings suffered the most serious damage among these four building types. A large number of school buildings were mainly brick houses, with large rooms and corridors, offering very poor seismic resistance. Schools in poor areas have small budgets; unlike schools in the cities, they cannot collect huge fees and so funds are scarce. When new building standards for earthquake resistance are issued, government departments, major businesses, etc. all appraise and reinforce their buildings; rural schools do not have funds to do this. Executive vice governor Wei Hong confirmed on 21 November 2008, that 19,065 schoolchildren had died, and 1300 schools needed to be reconstructed, with an initial relocation of 25 townships, including Beichuan and Wenchuan.

Figure 1.26 Yingxiu middle school, Yingxiu Town in Wenchuan County. The lowest storey collapsed completely and walls show many X shape cracks.

Figure 1.27 Damage to factory buildings at Dongfang Steam Turbine Works (DFST) in Mianzhu City is representative. Most buildings used traditional bent columns. The precast truss, and ribbed roof slab collapsed entirely, causing serious losses of life and facilities. Fortunately, seven similar buildings survived the earthquake thanks to the lightweight colored sandwich steel plate or corrugated steel replacing the precast ribbed roof slab a few years ago. Factory buildings with steel trusses remained basically intact thanks to their satisfactory bracing system, despite using a heavy precast ribbed roof slab. Factory buildings using a portal frame withstood the earthquake well.

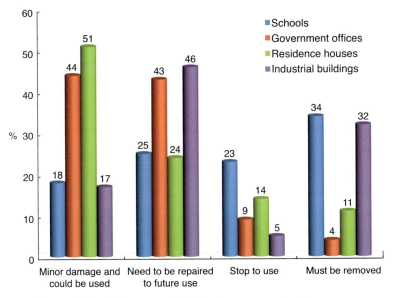

Figure 1.28 Histogram of building damage according to use.

In the aftermath of the earthquake, the Chinese government decided that the design standard for new schools and rebuilt schools should correspond to one degree of intensity higher than the national standard for the place where each school is located.

Local industrial plants in towns and villages are mostly masonry structures. Since seismic design requirements were very low, this led to serious damage. Multi-frame structures constructed for government agencies suffered the lightest damage.

Lessons learned

Many lessons were learned from the building damage survey.

First, the probable maximum intensity defined in the 2001 code for the Longmenshan fault zone area is now recognized to have been too low. We knew the existence of the Longmenshan fault zone since the 1920s, and we also knew of many historical large earthquakes in the central segment of the China's North-South seismic belt, and so it is hard to understand why the intensity of the Longmenshan fault zone was set too low on the new version seismic zoning map. Perhaps it was done in recognition of recent economic development; however scientific seismic hazard assessment should be independent of politics as well as the economy.

The surveyed (actual) intensities of most areas hit by the Wenchuan earthquake significantly exceeded the intensities on the zoning map. The expert group investigated building damage in five cities, where the scale differences between map intensity and actual intensity ranged from 1 to 2 degrees.

However, even though design intensity was low, the presence of a design standard proved to be vitally important. Many buildings that were built to the standard did not collapse even in the hardest hit areas. Buildings that incorporated the seismic design on average suffered much less than those that did not. Buildings that did not have any or proper seismic design include not only old buildings but also many newer buildings that were built when the 1989 or 2001 seismic design code was already in place.

Second, the severity of damage is related closely to the level of economic development of a region (Figures 1.29, 1.30). The economical developments of Mianyang and Dujiangyan are much better than Jiangyou and Anxian, therefore, the ratio of collapsed buildings in Mianyang and Dujiangyan are much less than Jiangyou and Anxian (Table 1.7). Statistics of building collapse

Figure 1.29 Mianyang gymnasiums, experienced intensity VIII during the Wenchuan earthquake. Most stadiums/gymnasiums in earthquake-affected areas were space frame structures (space truss and lattice shell). Featuring a light-weight roof and superior integrity, the space frame structure demonstrated its superior earthquake resistance capacity. In the Wenchuan earthquake some stadiums and gymnasiums stayed intact and were used as shelters for earthquake victims (photo by Chen Qinggang, Hangzhou Daily).

Figure 1.30 Rural residences, consisting mostly one- or two-storey masonry buildings, suffered much more extensive collapse than urban residences.

for Chengdu City and the five hardest hit prefectures (Mianyang, Jiangyou, Dujiangyan, Anxian, and Mianzhu) was reproduced in Table 1.8 after some regrouping (Tian et al., 2008). The absolute value of each collapsed fraction in this table is of limited usefulness because most of the areas involved, away from the Longmenshan fault zone, suffered seismic intensity VII or less and are not expected to have had extensive building collapse. It is the comparison

Table 1.7 Damage statistics for five cities where intensity was underestimated by 1–2 degrees (data from Civil and Structural Groups of Tsinghua University et al., 2008; modified)[*]

Map intensity	Mianyang	Jiangyou	Dujiangyan	Anxian	Mianzhu
GB500011-2001	VI	VII	VII	VII	VII
Actual intensity	VIII	VIII	IX	VIII–IX	VIII
Can be used	39(64%)	20(17%)	46(48%)	20(53%)	7(26%)
To be repaired for further use	9(15%)	63(53%)	32(34%)	6(16%)	5(19%)
Use prohibited	11(18%)	7(6%)	9(9%)	3(8%)	0(0%)
Demolition essential	8(3%)	30(25%)	8(8%)	9(24%)	15(56%)

* 384 buildings were investigated.

Table 1.8 Statistics of building collapse caused by the Wenchuan earthquake for Chengdu and five worst-hit prefectures (data after Tian et al., 2008)

Building type	Total construction area ($\times 10^6$ m^2)	Area of collapsed buildings ($\times 10^6$ m^2)	Collapse fraction (%)
Urban residential buildings	233.3048	4.1108	1.76
Rural residential buildings	167.1414	8.0814	4.84
Primary and secondary schools	19.3036	0.9214	4.77
University and trade schools	12.8002	0.1818	1.42
Hospitals	3.9480	0.1664	4.22
Local health clinics	2.6717	0.2970	11.09
Community centers and libraries	0.4242	0.0606	14.29
Gymnasiums, sport facilities	0.1988	0.0090	4.53
Government office buildings	17.6485	1.1380	6.45
Business/Organization office buildings	22.8281	0.7685	3.37
Shopping centers, restaurants, etc.	15.6982	0.9515	6.12
All types of shipping storage	3.8991	0.0551	1.41
Factories, plants	59.9593	8.2921	13.83
Industrial storage buildings	10.8089	0.3843	3.56

between collapsed fractions of different building types that is revealing. Factories, plants, community centers, public libraries, and local health clinics have the poorest performance. Fortunately, some of these buildings were not very densely occupied. The best performing buildings are those of shipping storage, universities, trade schools, and urban residences.

The large difference in living standards between the urban and rural environment exists at the present day in China. Rural residences, consisting of mostly one-storey or two-storey masonry buildings, suffered much more extensive collapse than urban residences (Table 1.8). The reason is that there was generally no seismic design for rural housing. Although the design code, mandatory by law, is not limited to the urban area, historically there has never been a mechanism to enforce it in the countryside. A survey conducted in 2003 in a few villages within the administrative areas of Mianzhu and Shifang Cities (see Figure 1.3 for city locations) showed that 98% of the houses inspected had no formal engineering design, and in 97% no protection measures against earthquakes had been taken (Zhang et al., 2009).

China is one of the countries which suffers the most severe earthquake disasters in the world, and the solution to earthquake prediction is still universally recognized as far off. Therefore, it is of great significance to summarize and draw together conscientiously and comprehensively the experience and lessons from earthquake disasters and building damage. This must be done in order to upgrade the level of capital construction in China, enhance the seismic resistance capacity of buildings, and reduce casualties and property damage as much as possible when an earthquake occurs.

1.4 Earthquake damage to lifelines

We summarize below the damage to lifelines including transportation (roads and railroads), tele-communications, and electricity and water supply systems.

Transportation

Wenchuan is located in a mountainous area, where roads are the most important resource for transportation. Roads in the Wenchuan area can be categorized into three types: national, county and rural roads. The Wenchuan earthquake caused damage to all the local national and county roads. Most of the destruction occurred in the zones of high earthquake intensity and where there were severe landslides and mudflows.

Nearly all the roads in the hardest-hit areas of Sichuan were damaged. After the earthquake, roads to Wenchuan, Maoxian, Beichuan, Qingchuan, Pingwu and the other hardest hit areas were completely cut off, making relief operations in the affected areas very difficult. Three months after the earthquake, three main roads had still not fully been opened: national road 213 (Yingxiu to Wenchuan), and provincial roads 303 (Yingxiu to Gengda) and 302 (Maoxian to Beichuan) (Zhuang et al., 2009).

We first examine the destruction of the road bridges (Figure 1.31).

There are numerous road bridges in the Wenchuan area. According to collected statistics for 1657 road bridges within zones of seismic intensity greater or equal to VII (Zhuang et al., 2009), most bridges showed good earthquake resistance capability, since only 1.09% of these bridges were destroyed (Table 1.9).

Besides the road bridges (particularly the national and provincial road bridges), road tunnels constructed in recent years showed good earthquake

Figure 1.31   (a) Yingxiu bridge (Wenchuan County, intensity X), where a piece of bridge deck collapsed. Note: the rest stayed intact, with no damage (courtesy of Lu Ming). (b) The most severe damage to a highway bridge concerned the Baihua bridge collapse on the Dujiangyan to Yingxiu section of national road 213 (from "renmin" website).

resistance capability; although the Wenchuan earthquake shaking was very strong, damage to most tunnels was very minor (Figure 1.32).

We note that in the mountainous Wenchuan area, many new tunnels and road bridges had recently been constructed on national and provincial roads, with strict building codes and high technology being applied. In the strong earthquake shaking, no tunnels collapsed and only a few of these bridges collapsed, though many bridge columns were damaged. Thus bridge damage was not the main reason for traffic disruption. Traffic disruption was mainly caused by landslides as well as damage to county and rural bridges. Roads constructed on mountain slopes suffered major damage through large and numerous landslides, causing widespread and severe traffic disruption (Figure 1.33). This demonstrates the advantage of using bridges and tunnels to traverse mountainous terrain.

Figure 1.32 Damage at Longdongzi tunnel near Dujiangyan City, intensity X. The tunnel was visible damage to the film all the scenes. It is extremely rare during Wenchuan earthquake that fault through the tunnel, therefore, most of the tunnel (even in X degrees area) was slightly damaged, or no damage during earthquakes (courtesy of Lu Ming).

Figure 1.33 Mountain landslide and debris flow, resulting in significant road damage and traffic disruption due to collapse of the road foundations.

Table 1.9 Description of damage to road bridges (Zhuang et al., 2009)

Type	Number	No damage or light damage (%)	Moderate damage (%)	Heavy damage (%)	Destroyed (%)
Highways	576	91.49	5.21	2.60	0.35
National or provincial roads	1081	78.17	11.01	8.78	1.48
Total	1657	82.81	8.99	6.63	1.09

Other national roads, and county roads, suffered much more serious damage from the earthquake because the standards of design were lower. County roads are numerous and widely distributed, so the damage had a great impact on traffic (Figures 1.34, 1.35).

In the rural areas, many old bridges, mainly stone bridges, suffered serious damage under the strong shaking. Although damage to rural roads was widespread, so that almost every village was affected, the scale of damage was limited and repair was easier.

Figure 1.34 Pengzhou, Xiaoyudong bridge, bridge collapse of two cross (courtesy of Lu Ming).

Figure 1.35 Under strong shaking in rural areas, many of the old stone bridges suffered serious damage. Collapse of stone arch bridges in (a) Beichuan (Chenjiaba), (b) Bailu Village, Pengzhou.

Railway systems

A wide network of national railways crisscrosses the Sichuan region, and there are many local railways serving mines and industrial plants. When the earthquake struck, 149 freight trains and 31 passenger trains were in service, but no passenger was injured.

Three national railways run through the earthquake epicenter region: Baoji–Chengdu, Chengdu–Kunming and Cheng–Yu. The most common rail accidents were caused by landslides. There were 4 landslides on the Baoji–Chengdu line, 4 landslides on the Chengdu–Kunming line, and 7 landslides on the Cheng–Yu line. The earthquake also caused communication system power interruption at 34 stations on the Baoji–Chengdu line.

The most serious accident occurred at No. 109 tunnel on the Baoji–Chengdu line at 14:28 on 12 May (Figure 1.36). Large rocks dislodged from the mountain by the Wenchuan earthquake derailed a train of 40 trucks and blocked the tunnel. The train was carrying 500 tons of aviation gasoline, which spilled and ignited; the fire caused severe damage to railway facilities. After continuous rescue work over eight days, all the trucks were pulled out.

Damage to rails involved buckling and large deformation either in

Figure 1.36 The most serious railway accident occurred at No. 109 tunnel on the Baoji–Chengdu line on 12 May. Large rocks falling from the mountain derailed a train of 40 gasoline trucks, blocked the tunnel, and caused an oilspill fire.

a vertical or horizontal direction (Figure 1.37). The extent of the damage principally depended on the track foundations (Figure 1.38). In general, if tracks were on saturated loose sand or saturated alluvium, then serious damage could be induced by ground subsidence. In contrast, rail tracks within the intensity XI zone were relatively undeformed since the soil was hard and the water table was deep.

Figure 1.37 Damage to rails involved buckling and large deformation in either a vertical or horizontal direction. (a) Baoji–Chengdu railway line, Shifang section. (b) Railway line deformed by landslide, Shifang.

Figure 1.38 Strong ground shaking and deformed tracks caused train derailment.

Electricity, water supply, sewage and telecommunication

All engineered lifeline systems—electricity, water supply, sewage and telecommunications were crippled by the $M_S8.0$ earthquake (Figures 1.39–1.43). Three 500kV-electricity transmission lines and 56 220kV-lines tripped after the earthquake and 122 110kV-lines, 110 35kV-lines and 795 10kV-lines suffered outages. 21 hydroelectric plants and four thermal power plants and power system were damaged. A total of 2.46 million users suffered power outages. The disruption of the power grid was mainly due to fallen pylons, damage to transformers, circuit breakers and other high voltage equipment, and local damage due to broken poles, and breakdowns of village transformers.

Immediately after the earthquake there was widespread failure of the water supply in the stricken area because electricity was not available for pumps and other equipment. Meantime, severe damage occurred to water pipes, water towers, etc. According to the Ministry of Water Resources report, 8426 waterworks and 47,642 km of pipeline were damaged.

Dujiangyan is home of the Dujiangyan Irrigation system, an ancient water diversion project which is still in use and is a UNESCO World

Figure 1.39   High voltage transmission line pylons toppled, Yingxiu Town in Wenchuan County.

Figure 1.40   Damage to the middle part of the brick water tower in Shifang. In the earthquake hit areas, chimneys and water towers were built of either brick masonry or reinforced concrete. The earthquake resistance capability of brick buildings is much less than those built of reinforced concrete. Damage to brick chimneys is mainly at the top, but with increasing earthquake intensity, the zone of damage moves downwards. The worst damage to this brick water tower is close to its base.

Figure 1.41　Jiangyou Two-horse Group factory. Of five chimneys, four were decapitated.

Figure 1.42　Destruction of mobile communication station in Wudu, Gansu Province, 450 km far from epicenter.

Figure 1.43 A communications tower destroyed at Wenchuan.

Heritage Site. Its famous Fish Mouth was cracked but otherwise the project did not suffer significant damage.

China Mobile had more than 2300 base stations out of action due to power disruption or severe telecommunication traffic congestion. Half the wireless communications were lost in Sichuan Province. China Unicom's service in Wenchuan and four nearby counties was cut off, with more than 700 masts inoperative. Damage to communication systems included: damage to base station masts or housing, broken frame poles, mast communications equipment tilted, displaced or destroyed; TV broadcast cables disconnected.

"Quake lakes"

Many rivers are lifelines for humanity. In this sense, it can be said that a barrier to river flow represents damage to a lifeline.

Within days of the magnitude 8.0 earthquake that shook the Sichuan Basin, flooding became a hazard. As a result of the earthquake and its many strong aftershocks, many rivers became blocked by large landslides, resulting in the formation of "quake lakes" behind the blockages. Massive amounts of water pooled up at a very high rate behind these natural dams and it was feared that the blockages would eventually crumble under the ever-increasing water pressure, endangering the lives of millions of people living downstream. By 19 May 2008, 21 lakes had formed throughout the basin and it was estimated that half of them were still of potential danger to local people. Entire villages had to be evacuated because of the resultant flooding upstream from landslides.

Earthquake-created dams present a dual danger. Apart from the upstream flooding that occurs as a lake builds behind the natural dam, the piles of rubble that form the dam may be unstable. Another quake or even the water pressure behind it could burst the dam, sending a wall of water downstream. Downstream floods can also occur when water begins to cascade over the top of the dam. Thousands of people were evacuated from Beichuan on 17 May when one such lake threatened to burst.

The most precarious of these quake-lakes was the one located in extremely difficult terrain at Tangjiashan mountain, accessible only by foot or by air. The lake formed in a mountain valley upstream of the small city of Beichuan (Figures 1.44, 1.45). Two Mi-26T heavy lift helicopters were used

Chapter 1 · The Wenchuan earthquake   053

Figure 1.44   Tangjiashan earthquake lake (Yansai lake) created in the upper Qianjiang river, potentially endangering the lives of millions of people living downstream (Beichuan County: 3.5 km, Mianyang City: 30 km) (photo by Zhu Wei, Xinhua News Agency).

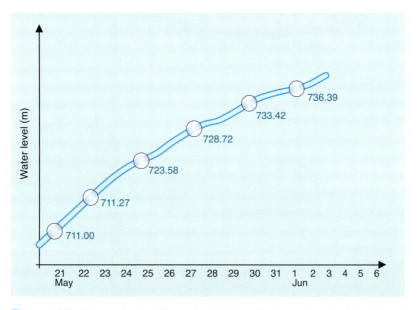

Figure 1.45   Time variation of water level of Tangjiashan earthquake lake (Yansai lake). A massive amount of water was accumulating at a very high rate behind the natural landslide dam.

to bring earthmoving tractors to the affected location (Figure 1.46). Over 1200 engineers and workmen, explosive specialists and other personnel took part in this work. Five tons of fuel to operate the machinery were airlifted to the site, where a sluice was constructed to allow the safe discharge of the accumulating water. Downstream, more than 200,000 people were evacuated from Mianyang by 1 June in anticipation of the dam bursting.

After further treatment, many of these earthquake lakes have become reservoirs.

Figure 1.46 This Mi-26T heavy lift helicopter was used to bring heavy (13 tonnes) earthmoving tractors to the affected location (photo by Cai Yang, Xinhua News Agency).

1.5 Secondary disasters: landslides and rock falls

Secondary disaster areas are mainly situated in the transitional area from the Tibetan Plateau to the Sichuan Basin with the Longmen Mountains as the dividing line, where the geological and topographical features between west and the east are quite different from each other, as is their socioeconomic development level. In general, these areas have the following characteristics:

1. The disaster areas are characterized by complex topographic and climate conditions, with varied distributions of plains, hills, plateaus, mountains, etc. Certain areas feature large differences in relative altitude and associated vertical changes in climatic conditions. These areas have typical mountain canyon topography.
2. Natural disasters haunt these mountainous plateau areas, where seismic fault zones crisscross and the probability of earthquakes is relatively high; sites for potential geological hazards like landslides, mud flows and rock falls are densely and widely distributed across these areas, and pose a grave threat.

Earthquakes can produce slope instability leading to landslides, which are a major geological hazard. The landslide danger may persist while emergency personnel are attempting rescue.

Deaths from quake-triggered geological disasters (landslides, rock falls, debris flow) accounted for over a third of the total Wenchuan earthquake deaths, which is extremely rare in earthquake disaster history.

We now discuss the effect of geomorphology on the damage, given that the Wenchuan 8.0 earthquake induced giant landslides and mud flows.

Geomorphology of Wenchuan region

Wenchuan earthquake occurred on the Longmen Mountain Fault System (LMFS) with 250 km long ground surface raptures. The LMFS is a series of faults striking in a northeast direction, on a North-South zone of high topographical and geophysical gradients between the Tibet Plateau on its western side and the Yangzi Platform on its eastern side. This gradient belt, named the North-South seismic belt (see section 2.3), crosses the

LMFS and divides Earth's crust in China into eastern and western parts. There are high mountains and deep valleys in the LMFS. Within 50 km from east to west, the average elevation changes sharply from 2.5 km to 5.5 km, with the highest peak at 7.5 km (Figure 1.47). Towards the core, the thickness of the crust also changes sharply from 40 km to 60 km.

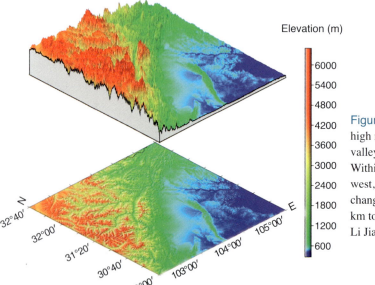

Figure 1.47 There are high mountains and deep valleys in the LMFS. Within 50 km from east to west, the average elevation changes sharply from 2.5 km to 5.5 km (courtesy of Li Jiancheng).

Figure 1.48 Landslides near Yingxiu Town, Wenchuan County. There are many landslides along the 5 km stretch (courtesy of Guo Huadong).

Mountain regions are already a fragile ecological environment. The Wenchuan earthquake triggered a large number of secondary geological disasters (Figure 1.48). There were at least 20 giant landslides, and each of them caused more than 100 deaths.

It is estimated that there were over 50,000 earthquake triggered landslides and avalanches, among them 30 giant landslides with volumes of over 10 million m^3. The largest landslide occurred in Anxian County with a volume of about 750 million m^3, forming a 550 m high rockfall dam. The Anxian landslide is one of the largest landslides in the world. Many landslides destroyed houses and buildings, interrupted traffic, hampered rescue efforts and caused heavy casualties, as well as changing the colour of their mountains (Figure 1.49). One landslide in Beichuan County entombed more than 1600 people (Figure 1.50). Secondary geological disasters often cause direct damage to infrastructure and pose a serious threat, besides causing tremendous difficulties in hampering the smooth operation of post-disaster relief and reconstruction work.

Chapter 1 · The Wenchuan earthquake | 059

Figure 1.49 Large scale landslides in Wenchuan County changed the color of the mountains.

Figure 1.50 Following the Wenchuan earthquake, most deaths caused by landslides were caused by the Wangjiayan landslide. Two giant landslides in Beichuan County: Wangjiayan landslide (on the left bank of Qian river) and Jingjiashan landslide (on the right bank of Qian river). The Wangjiayan landslide directly buried more than 1600 people. The atmospheric air wave generated by the landslide destroyed all houses within 100 m of the front of the landslide (courtesy of Guo Huadong).

Giant landslides

Large volumes of material can move downslope under the pull of gravity, and some do so catastrophically (Figures 1.51, 1.52). These mass movements usually occur in the rainy season. Water plays many important roles in mass movements, both externally and internally. Rainfall is an external factor, since it comes from the atmosphere. Runoff of rainwater causes external erosion that sets masses moving on slopes, and it undercuts bases of the slopes, causing hillsides to fail and move. Internally, the roles of water are to weaken rock and soil through increased pore pressure, and to reduce frictional forces preventing mass movement.

May is the dry season in Wenchuan area; before and during the earthquake there was no rainfall, thus water could not play a significant role. However rock mass movements can also occur in dry conditions.

Beichuan middle school in Beichuan County was sited on a small col between two mountains, and more than 2900 students studied there. In the afternoon of 12 May two five-story classroom buildings collapsed as a result

Figure 1.51 The Jingjiashan landslide buried Beichuan middle school completely, and 600 students and teachers lost their lives. The whole school lies in ruins, with the flag in front of the school still flying (courtesy of Huang Runqiu).

Figure 1.52 The biggest landslide caused by the Wenchuan earthquake is the Daguangbao landslide in Anxian County, for which the sliding distance is 4.5 km, the accumulation body width 2.2 km, landslide debris flow area 10 km^2, and the estimated volume is 750 million m^3. Daguangbao landslide is not only the largest known landslide within China so far, but also the one of the world's known largest landslides with a volume more than 500 million m^3 (courtesy of Xu Qiang).

of the earthquake and buried those within, killing 700 people.

A Beichuan student who was rescued gave the following description of the scene at the time: "There were no precursors. 12 May was hot and only slightly gloomy, weather we had long taken for granted. As usual after the midday nap there was a rush to the classrooms. In the new building of the multimedia classroom art class, the teacher gave us a picture to look at, and about ten minutes later the earthquake started, we all understand this is no longer a joke, and so rapidly we dived under our desks. Some classmates suddenly panicked, screaming, crying…".

Mud flows

Flows are mass movements that behave like fluids. The materials within the flows can be massive boulders, sand, clay, ice or mixtures of them all. There is a complete gradation from debris slides that move on top of slip surfaces, to debris flows that do not require any basal slip surface. Many names are used to describe such flows, such as, earthflow, mudflow, debris flow, and debris avalanche. Many flows occurred due to the Wenchuan earthquake—all were characterized by fluid-like behavior, but none were of water. Figures 1.53 & 1.54 show the mud flows at Yingchanggou.

Figure 1.53 Mud flows at Yingchanggou, Pengzhou. Mud flows can devastate both villages and countrysides if they are large enough.

Figure 1.54 Mud slide at Yingchanggou, Pengzhou. A mudslide is the most rapid and fluid type of downhill mass movement.

Rock falls

Rock falls occur when elevated rock masses separate along joints, rock layers and other planes of weakness. When a mass detaches, usually it falls downward through the air in freefall, then after hitting the ground, it moves by bounding and rolling. Falls may be triggered by heavy rainfall, frost wedging, earthquakes, volcanic eruptions, etc. The Wenchuan earthquake occurred in May, which is not a rainy season in Sichuan Province. There was no rainfall, no frost wedging (and no volcanic eruption!) at that time. Thus we deduce that the large scale rock falls were caused by the Wenchuan earthquake (Figure 1.55).

Rock masses are commonly fractured, with many joint directions. Each fracture is a weakness that separates blocks of rock. Careful inspection of the fallen rock masses reveals old weakness surfaces caused by erosion, usually by weathering.

Some rock blocks which were thrown down from the mountains showed fresh weakness surfaces. Acceleration seismographs located on mountainsides show that the PGA exceeded 1 g up to 20–30 km from the epicenter, and since the PGA at the top of hills can be greater, some intact rock may be fractured and thrown down from the hills (Figure 1.56). This is the reason why large fallen rocks can be found downslope with fresh weakness surfaces (Figure 1.57).

The severe secondary disasters (landslides, mud flows, rock falls, etc.) caused by the Wenchuan earthquake are a reminder that hazard assessment and land-use regulations are the most important issues in mountainous areas for disaster mitigation. Adoption of effective land-use regulations and

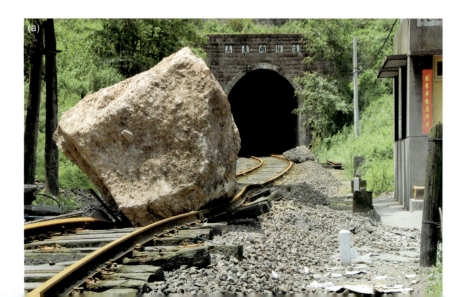

Chapter 1 · The Wenchuan earthquake 067

Figure 1.55 Rock falls.
(a) Beichuan County;
(b) Anxian County;
(c) Liuba Village of Hanzhong City, Gansu Province;
(d) bus crash at rockfall.

building codes must be based on scientific research. Land-use regulations will discourage new construction or development in identified hazard areas without first implementing appropriate remedial measures. Attention must also be paid to structural and non-structural mitigation measures, whereby

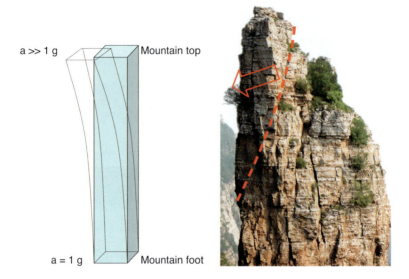

Figure 1.56 some parts of mountain tops were thrown out under strong earthquake shaking (PGA >> 1 g) by an action similar to a catapult.

Figure 1.57 Rock fallen from mountain top, fresh rupture surface on it (courtesy of Huang Runqiu).

remedial techniques (such as buttresses, soil reinforcement, retaining walls, etc.) are applied in respect of existing landslides in close proximity to public structures.

References

Chen Q F, Wang K L. 2010. The 2008 Wenchuan earthquake and earthquake prediction in China. Bull Seismol Soc Amer, 100(5B): 2840–2857.

Chen Y, Chen Q F, Liu J, et al. 2002. Seismic Hazard and Risk Analysis: A Simplified Approach. Beijing: Science Press.

China Earthquake Administration. 2008. Report on Strong Earthquake Motion Records in China, vol 20, No 1. Beijing: Seismological Press. (in Chinese)

Civil and Structural Groups of Tsinghua University, Southwest Jiaotong University, Beijing Jiaotong University. 2008. Analysis on seismic damage of buildings in the Wenchuan earthquake. J Build Struct, 29(4): 1–9. (in Chinese)

Li X J, Zhou Z H, Huang M, et al. 2008. Preliminary analysis of strong-motion recordings from the magnitude 8.0 Wenchuan, China, earthquake of 12 May 2008. Seismol Res Lett, 79(6): 844–854.

Liu H X. 1986. The Damage of the Tangshan Earthquake, vol.4. Beijing: Seismological Press. (in Chinese)

Liu Z, Wu Z L. 2005. A simple model of reported casualties during earthquakes and earthquake-generated tsunamis. Earthquake Res China, 21 (4): 526–529. (in Chinese)

Ministry of Construction of the People's Republic of China. 2001.GB50011-2001, Code for Seismic Design of Buildings. Beijing: China architecture & Building Press.(in Chinese)

Science and Technology Expert Group of the Ministry of Science and Technology. 2008. Wenchuan earthquake hazard analysis and comprehensive evaluation. Beijing: Science Press. (in Chinese)

Tian W, Chen Z, Xiang X, et al. 2008. Analysis of and reflections on building damage caused by the 12 May Wenchuan earthquake in Sichuan. Sichuan Constr Sci, 34(6): 123–129. (in Chinese)

Wen X Z, Ma S L, Xu X W, et al. 2008. Historical pattern and behavior of earthquake ruptures along the eastern boundary of the Sichuan–Yunnan faulted-block, southwestern China. Phys Earth Planet In, 168(1/2):16–36.

Wu X Y, Gu J H, Wu H Y. 2009. A modified exponential model for reported casualties during earthquakes. Acta Seismol Sin, 31 (4): 457–463,477. (in Chinese)

Xu X W, Wen X Z, Ye J Q, et al. 2008. The M_S8.0 Wenchuan earthquake surface ruptures and its seismogenic structure. Seismol Geol, 30(3): 597–629. (in Chinese)

Zhang H, Xiang D, Hu Y. 2009. Earthquake-safe Rural Housing Project in Deyang City passed the test of the great Wenchuan earthquake. Cities and Hazard Mitigation, 3: 25–28. (in Chinese)

Zhou Z H. 2008. The strong ground motion recordings of the M_S8.0 Wenchuan earthquake in Sichuan Province. Earthquake Res Sichuan, 129(4): 25–29. (in Chinese)

Zhuang W L, Liu Z Y, Jiang J S. 2009. Analysis of road bridges after the '5.12' Wenchuan earthquake. Highway (5): 130–139. (in Chinese)

Figure 2.1 Clock of Hanwang Village, Sichuan Province, China, stopped at 14:28 on 12 May 2008, the occurrence time of the great Wenchuan $M_s 8.0$ earthquake (courtesy of Lu Ming).

Seismological features 2

072 / Seismic source parameters for the mainshock
084 / Historical earthquakes and aftershocks
093 / Geophysical investigations before the earthquake
104 / Tectonic setting of Wenchuan earthquake
115 / Seismic waves generated by earthquake

The 2008 Wenchuan earthquake was a deadly earthquake of magnitude $M_S 8.0$, which occurred at 14:28 on 12 May 2008, in Sichuan Province of China and killed at least 69,226 people, with 17,923 missing. Approximately 15 million people lived in the affected area, and it left about 4.8 million people homeless. It was the deadliest earthquake to hit China since the 1976 Tangshan earthquake, which killed at least 240,000 people, and the strongest in China since the 1950 Chayu earthquake, which registered 8.5 on the Richter magnitude scale.

It is known as the Wenchuan earthquake, named after the location of the earthquake's epicenter, Wenchuan County in Sichuan Province. The epicenter was 70 km west-northwest of Chengdu, the capital of Sichuan Province. The earthquake was also felt in nearby countries and as far away as both Beijing (1500 km away) and Shanghai (1700 km away) where office buildings swayed with the tremor.

Direct economic losses of Wenchuan earthquake were 845.1 billion yuan. On 6 November 2008, the central government announced that it would spend 1 trillion yuan (about 146.5 billion dollars) over the next three years to rebuild areas ravaged by the earthquake.

We shall summarize here the seismological and geological observations on the Wenchuan earthquake which provide clues as to why this magnitude 8.0 earthquake was so destructive.

Detailed seismological studies and geological investigations were performed in Wenchuan area after the main shock. The key elements of these scientific results have been summarized in a book entitled *Scientific Research Report of Wenchuan M8.0 Earthquake* (2009), and prepared by a group of 23 specialists gathered together by the China Earthquake Administration(CEA). Unless otherwise stated, most of the scientific data discussed in this chapter are from this comprehensive and definitive publication. We emphasize the following aspects which are particularly relevant to the discussion: the source parameters of the main shock, spatial and temporal distribution of the aftershocks, historical records of earthquakes in Wenchuan region, geophysical background surveys, and regional geology.

2.1 Seismic source parameters for the mainshock

Source parameters

The principal source parameters of the Wenchuan main shock were determined by the China Earthquake Network Center (CENC) as follows:
- Origin time: 2008-05-12 14:28:04 (Figure 2.1)
- Epicenter: 30°57' N, 103°24' E
- Focal depth: 16 km (Huang et al., 2008)
- Magnitude[*]: $M_S 8.0$

The source parameters determined by various international seismo-

---

[*] Immediately after the Wenchuan earthquake, the preliminary magnitude was reported as 7.8; this preliminary determination was revised after more through analysis of the full records, and the revised magnitude $M_S 8.0$ was issued later.

logical institutions worldwide are in reasonable agreement with those determined from the Chinese data, but there were minor differences in magnitude determinations (Table 2.1):

Table 2.1 Magnitude values determined by major seismological institutions

Institution	Magnitude
CENC	M_S8.0
NEIC	M_S8.1
	M_W7.9
BGS	M_S8.0
	M_S 7.8
HRV	M_W7.9
CSEM	M_W7.8

CENC: China Earthquake Network Center.
NEIC: National Earthquake Information Center, U.S.A.
BGS: British Geological Survey.
HRV: Harvard University, U.S.A.
CSEM: European-Mediterranean Seismological Centre.

Since establishment of the World Wide Standardized Seismographic Network (WWSSN) in 1960's, and the later broadband Global Seismographic Network (GSN) in the 1980's, the Wenchuan earthquake has been the largest continental intraplate earthquake event recorded. Before the Wenchuan earthquake, the 1976 Tangshan earthquake measured by the WWSSN at M_S7.8 was the largest continental intraplate earthquake[*].

Focal mechanism

Seismologists refer to the direction of slip in an earthquake and the orientation of the fault on which it occurs as the focal mechanism. They use information from seismograms to calculate the focal mechanism and typically display it on maps as a "beach ball" symbol. This symbol is the projection on a horizontal plane of the lower half of an imaginary, spherical shell (focal sphere) surrounding the earthquake source (Figure 2.2). Orthogonal lines are scribed where two possible fault planes intersect the shell. The differently shaded zones indicate where the direction of first P motion is outward (grey) and inward (white). The stress-field orientation at the time of rupture governs the direction of slip on the fault plane, and the

---

[*] There have been two other great continental intraplate earthquakes after Tangshan earthquake with magnitudes close to 8: the Mani earthquake of 8 Nov 1997 with magnitudes 7.9 (M_S, NEIC) and 7.5 (M_W, NEIC), and the Kokoxili earthquake west of Kunlun Mountain on 14 Nov 2001 with magnitudes 8.0 (M_S, NEIC) and 7.8 (M_W, NEIC).

beach ball also depicts this stress orientation. In the schematic diagrams, the grey quadrants contain the tension axis (T), which corresponds to the minimum compressive stress direction, and the white quadrants contain the pressure axis (P), corresponding to the maximum compressive stress direction. Computed focal mechanisms show only the P and T axes and do not use shading.

The focal mechanisms are computed using a method that attempts to find the best fit to the directions of P-wave first motions observed at all seismometer stations. For a double-couple source mechanism (with only shear motion on the fault plane), compressional first-motions should lie only in the quadrant containing the tension axis, and dilatational first-motions should lie only in the quadrant containing the pressure axis. However, first-motion observations can often be in the wrong quadrant. This occurs because: (1) the algorithm assigned an incorrect first-motion direction because the signal was not impulsive; (2) the earthquake velocity model, and hence, the earthquake location is incorrect, so that the computed position of the first-motion observation on the focal sphere (or ray azimuth and angle of incidence with respect to vertical) is incorrect; or (3) the seismometer was mis-wired, so that "up" is "down". The latter explanation is unusual. For mechanisms computed using only first-motion directions, incorrect first-motion observations may greatly affect the computed focal mechanism parameters. Depending on the distribution and quality of first-motion data, more than one focal mechanism solution may fit the data equally well.

For mechanisms calculated from first-motion directions, as well as some methods that model waveforms, there is an ambiguity in distinguishing the fault plane on which slip occurred from the orthogonal, mathematically equivalent, auxiliary plane. We illustrate this ambiguity with four examples (Figure 2.2). The block diagrams adjacent to each focal mechanism illustrate the two possible types of fault motion that the focal mechanism could represent. Note that the view angle is 30° to the left of and above each diagram. The ambiguity may sometimes be resolved by comparing the two possible fault-plane orientations to the alignment of the foci of small earthquakes and aftershocks. The first three examples describe fault motion that is purely horizontal (strike slip) or vertical (normal or reverse). The oblique-reverse mechanism illustrates that slip may also have components of horizontal and vertical motion.

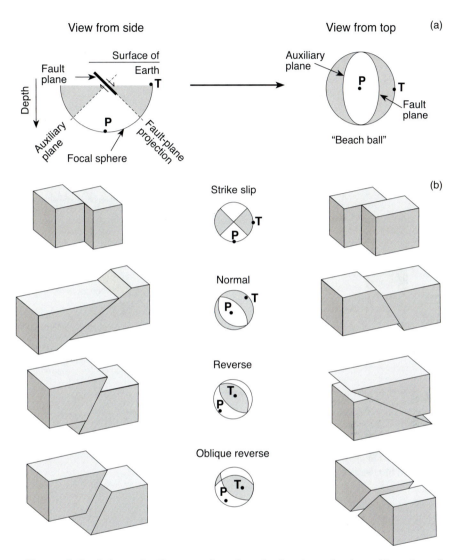

Figure 2.2 Schematic diagram of earthquake focal mechanisms (from http://earthquake.usgs.gov/learn/topics/beachball.php).

Focal mechanism solutions for the main shock were obtained by several research teams using seismic network data from China as well as teleseismic data from global networks. The focal mechanism solution is given in Figure 2.3.

Minor discrepancies exist among the different studies, but overall the solutions are in reasonable agreement. They all show one nodal plane striking NE–SW, and the other striking N–S. The P-axis (for principal compressive stress drop) is close to E–W, whereas the T-axis (for the

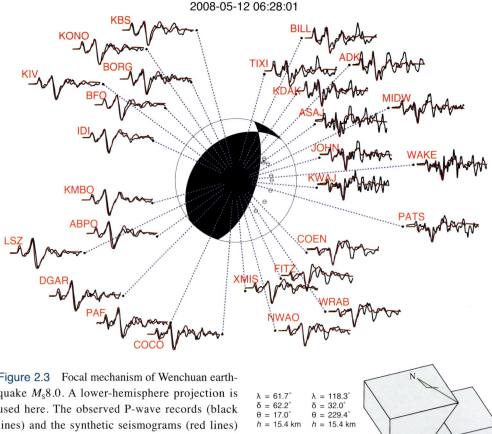

Figure 2.3 Focal mechanism of Wenchuan earthquake $M_s 8.0$. A lower-hemisphere projection is used here. The observed P-wave records (black lines) and the synthetic seismograms (red lines) based on the simple point model are compared. The green and blue circles with "+" and "−" indicate the polarity of the direct P waves, and the red circles are the projection of the stations. The parameters of two possible fault planes are listed, indicating the rake, dip, strike and source depth respectively. The derived source time function and schematic map of fault motion are plotted (from Wang et al., 2008).

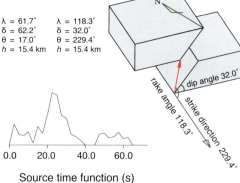

principal tensile stress drop) trends WSW–ENE. The spatial distribution of aftershocks indicates that the first nodal plane represents the fault plane for the magnitude 8.0 main shock.

The Longmenshan fault is a thrust fault which runs along the base of the Longmenshan Mountains in Sichuan Province, SW China. The strike of the fault plane is approximately NE. Motion on this fault is responsible for the uplift of the mountains relative to the lowlands of the Sichuan Basin to the east (Figure 2.4).

Chapter 2 · Seismological features | 077

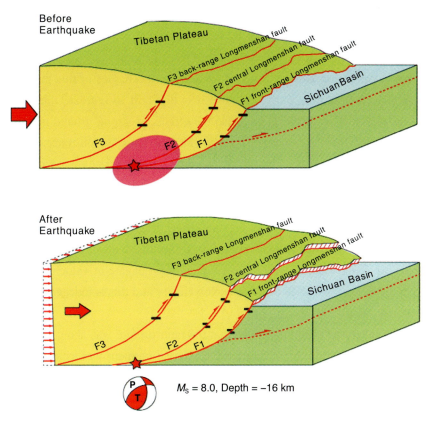

Figure 2.4 Earthquakes result from sudden motion on faults, which are usually oblique, with an upper part called the hanging wall, and the lower part the footwall. During the Wenchuan earthquake, the hanging wall moved upward and the footwall downward (thrust faulting). Motion occurred on three faults: F1, front-range Longmenshan fault; F2, central Longmenshan fault; F3, back-range Longmenshan fault.

Rupture process

The rupture process of the Wenchuan earthquake was quickly determined using long period seismic data from the Global Seismographic Network (GSN, Figure 2.5) a few hours after the occurrence of the main shock (http://www.cea-igp.ac.cn/special_issure/earthquake_situation/preliminary_result(1).pdf) and reported on the second day; this provided important information for rescue work in the field.

Figure 2.5 Epicenter (white star) of 2008 Wenchuan earthquake and spatial distribution of 21 long period seismic stations used for determination of the rupture process (from Zhang et al., 2008).

A tectonic earthquake begins by an initial rupture at a point on the fault surface, a process known as nucleation. Once the rupture has initiated it begins to propagate along the fault surface. The mechanics of this process are poorly understood, partly because it is difficult to recreate the high sliding velocities in a laboratory. Also the effects of strong ground motion make it very difficult to record information close to a nucleation zone.

Rupture propagation is generally inverted by using a fracture mechanics approach, likening the rupture to a propagating shear crack. The rupture velocity is a function of the fracture energy in the volume around the crack tip, increasing with decreasing fracture energy. The velocity of rupture propagation is an order of magnitude faster than the displacement velocity across the fault. Earthquake ruptures typically propagate at velocities that are in the range 70%–90% of the S-wave velocity and this is independent of earthquake size.

The inversion results indicated that the fault rupture was about 300 km long (Figure 2.6), and the released scalar seismic moment was estimated to be 9.4×10^{20}–2.0×10^{21} Nm. The slip distribution was highly inhomogeneous with an average slip of about 2.4 m (Figure 2.7). Four fault breaks broke the ground surface. Two of them were underneath Wenchuan–Yingxiu and Beichuan: the first was around the hypocenter (rupture initiation point),

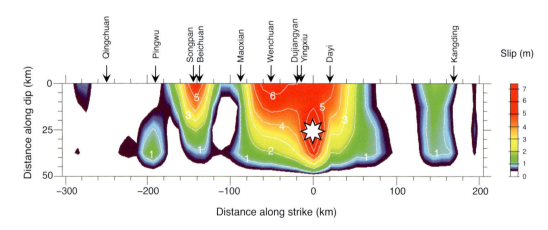

Figure 2.6 Static (final) slip distribution on the 2008 Wenchuan earthquake fault. The white star denotes the hypocenter (rupture initiation point). White lines are contours of the slip value (m). The arrows show the projected location of cities and counties on the fault trace (intersection of fault and ground surface) (from Zhang et al., 2008).

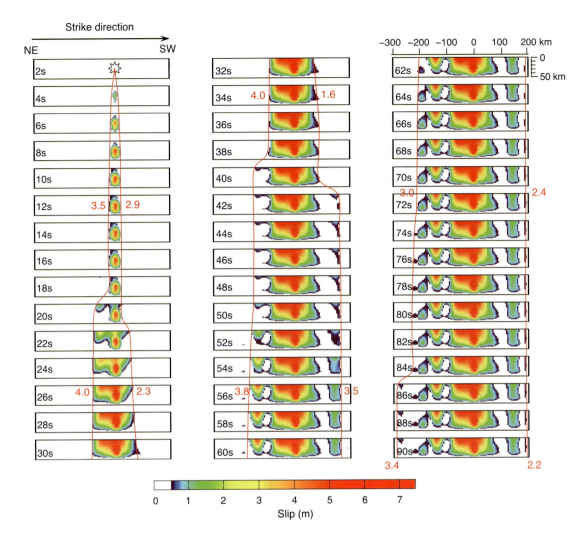

Figure 2.7 Snapshots of slip variation with time on the 2008 Wenchuan earthquake fault. The white star denotes the location of the hypocenter (rupture initiation point). Each rectangle denotes the fault plane slip distribution at the time indicated in its lower left corner. The earthquake rupture duration time was about 90 sec. The lowest rectangle on the right is the static (final) slip distribution at 90 sec. Red lines represent the evolution of rupture fronts with time. The rupture initiated at 15 km underneath the town of Yingxiu in Wenchuan County and stopped in Qingchuan County, northeast of the epicenter. The figures beside the red lines are the corresponding rupture velocity values (unit: km/s) (from Zhang et al., 2008).

where the largest slip was about 7.3 m; the second was underneath Beichuan and extended to Pingwu, with maximum slip about 5.6 m. The other two surface fault breaks were smaller, one having a maximum of 1.8 m and lying to the north of Kangding, the other having a maximum slip of 0.7 m and lying to the northeast of Qingchuan. Average and maximum stress drops over the whole fault plane are estimated to be 18 MPa and 53 MPa, respectively (Zhang et al., 2008).

Co-seismic deformation

A field survey after the mainshock discovered co-seismic deformation, which extended to almost 250 km along the N-E trend. Co-seismic deformation is responsible for the uplift of the mountains relative to the lowlands of the Sichuan Basin to the east (Figures 2.8, 2.9).

Figure 2.8 Simple thrust scarp. A 4 m vertical surface offset of the Wenchuan thrust fault at Hongkuo Village, Dujiangyan City, on the south part of the fault (courtesy of Xu Xiwei).

Figure 2.9 Thrust fault offset, showing clear scratches.

InSAR data provides information on coseismic deformation. Qiao Xuejun et al. (personal communication) explored 8 overlapping interferograms from the phased-array-type L-band SAR (PALSAR) of the Advanced Land Observing Satellite (ALOS) launched by the Japan Aerospace Exploration Agency (JAXA), because they provide a nearly complete map of ground deformation in the epicentral area. These SAR images were acquired from ascending paths 470–477 (Table 2.2), and each covers an approximately 70 km wide swath.

The radar images were processed using ROI_PAC software developed at JPL, with satellite orbits provided by JAXA. Interferograms were downsampled using a multi-look operation (4 looks in range and 20 looks in azimuth). A weighted power spectrum technique was applied to filter the fringes to produce the wrapped interferograms, each with a centre scene incidence angle of 34.3° and an azimuth of N12.8°W. The interferograms suffered severely from troposphere and ionosphere disturbances over the epicentral areas. Qiao et al. made no special effort to mitigate these effects on the interferograms, but the errors are relatively small (< 5 cm) compared to the deformation caused by a M8 earthquake.

The interferograms show no phase coherence in an elongated belt 30–40 km wide near the surface rupture, owing to distorted landscape and rugged terrain. Apart from the near-source region, coherence is excellent. These InSAR measurements show only a single component of the three-dimensional displacement field. So Qiao et al. projected the three components of the GPS LOS displacements on to the LOS (line-of-sight)

Table 2.2 Interferogram pairs of ALOS SAR images

No.	Path	Master image date	Slave image date	Perpendicular baseline (m)
1	470	2007-02-09	2009-06-29	−90
2	471	2008-02-29	2008-05-31	90
3	472	2007-01-28	2008-06-17	203
4	473	2008-02-17	2008-05-19	225
5	474	2008-03-05	2008-06-05	283
6	475	2007-06-20	2008-06-22	−34
7	476	2008-04-08	2008-05-24	−300
8	477	2008-04-25	2008-06-10	−70

direction and compared the LOS-directed GPS displacements to the InSAR range-offset measurements along several profiles normal to the surface rupture (Figure 2.10). These profiles show that the two data sets are in good agreement at a level of 3–4 cm on the Sichuan Basin and 6–8 cm on the Longmenshan, while they reveal the extent to which the InSAR data fails to capture the near fault deformation. Furthermore, the InSAR interferograms do not (visually) show the rotation of the displacement field that is evident in the GPS displacements. Because the LOS vectors are oblique to the rupture, the InSAR interferograms contain information about both the strike-slip and thrust-motion components of the slip distribution. Obviously the LOS vectors available to us do not uniquely describe the three-dimensional components of ground motion associated with this event.

It can be seen from InSAR data that the maximum relative displacement in the LOS direction (viewed from satellite) at the epicenter reaches 260 cm, and the overall vertical displacement between the two sides of the fault can be up to 330 cm.

The elastic rebound theory is an explanation for how energy is distributed during earthquakes. As plates on opposite sides of a fault are subjected to stress, they accumulate energy and slowly deform until their internal strength is exceeded. At that time, a sudden movement occurs along the fault, releasing the accumulated energy, and the rocks snap back to their original undeformed shape.

If a straight line is drawn over the fault just before an earthquake, the line will become a curved line after the earthquake occurs. The shape of the curved line is similar to the curve measured by GPS and InSAR (Figure 2.10). This means that the region of stress (energy) release for the Wenchuan earthquake must extend to at least 100 km on both sides of the Longmenshan fault.

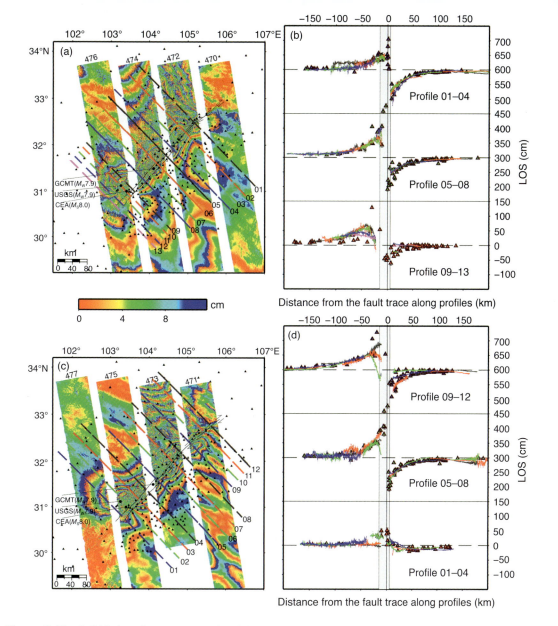

Figure 2.10 InSAR interferograms associated the Wenchuan earthquakes and profiles (courtesy of Qiao Xuejun).
The black triangles in b and d correspond to GPS sites shown in a and c. Black curved lines are surface ruptures from field investigations.
(a) InSAR fringes are derived from 4 ALOS tracks 470, 472, 474 and 476. The range of colors from blue to red, shown in the color bar at the bottom, corresponds to one fringe, representing ~ 11.8 cm of range change between the ALOS satellite and a point on the ground.
(b) GPS-derived surface displacements vs space-borne SAR range changes along profiles across the fault. InSAR data (colored dots) are subsampled along 100 meter-wide colored belts. Profiles 01–13 correspond to colored dashed lines arranged in order from southwest to northeast in a. GPS data (colored triangles) are extracted from the 80 km-wide profiles in Figure 2.10(a) and 3-dimensional displacements are projected onto the LOS direction.
(c) InSAR fringes from tracks 471, 473, 475 and 477.
(d) Profiles 01–12 correspond to the colored dashed lines placed in order from northeast to southwest in c (Qiao et al., personal communication).

2.2 Historical earthquakes and aftershocks

Historical earthquakes

In order to know the seismicity background from which the $M_S 8.0$ Wenchuan earthquake occurred, we first investigate the activity of local large and strongly felt earthquakes during the 2000 years before the Wenchuan earthquake. From the map of historical earthquakes with magnitude equal to or greater than 5.0 from 780 BC to 2007 (Figure 2.11), it can be seen that no $M > 7$ event had occurred in the middle segment of the North-South seismic belt (i.e. Longmenshan region) at least in the last 2000 years before 2008.

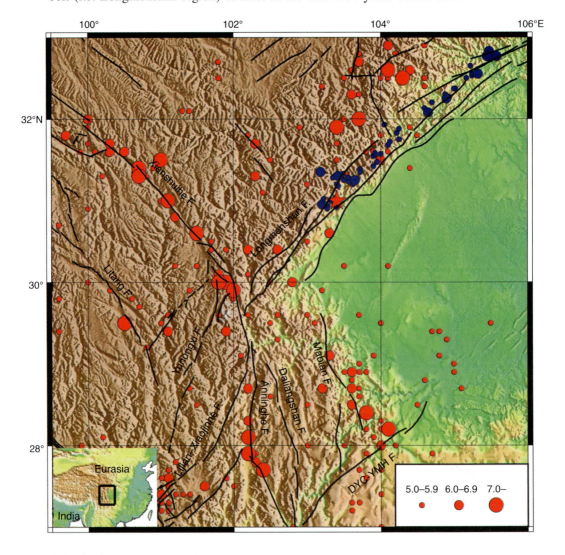

This suggests that relative to other active fault sections on the North-South seismic belt, a seismic gap had existed along the section of the Longmenshan fault zone for a long time.

We now consider "source areas" rather than earthquake epicenters. "Source areas" refer to the region of energy release—we can define them by using the intensity distribution map, as regions of intensity IX or above. In the west margin of Tibetan Plateau, most source areas have a well-defined ellipse. Figure 2.12 shows the distribution of source areas in the central segment of the North-South seismic belt. There are three main fault zones, forming a "Y" shape: Northwest—Xianshuihe fault; Northeast—Longmenshan fault; North-South—Anninghe fault. Figure 2.12 shows that most historical earthquakes occurred on the Xianshuihe and Anninghe faults, and no big earthquake occurred on the Longmenshan fault. Therefore before the Wenchuan earthquake, the seismic risk for the Longmenshan fault was ranked to be lower than that of the nearby Anninghe and Xianshuihe faults.

The earthquake monitoring network of Sichuan Province was updated in 1988, so that earthquakes with magnitude 2.5 or above could be recorded by the Sichuan network (Figure 2.13), as part of the Chinese digital seismic network. Thus we can investigate the earthquake catalogue over the last 20 years (1988–2008). During the 20 years before the Wenchuan earthquake, a quiescence of background seismicity emerged along the middle and south sections of the Longmenshan fault zone. The seismicity of the surrounding area remained at a normal level.

The size of the 2008 earthquake is much greater than that of the biggest historical event on the Longmenshan fault zone. This proves that potential seismic hazards along large-scale active fault zones with relatively low slip-rates can't be assessed correctly from historical earthquake records which are available only for several hundred or a few thousand years.

Figure 2.11 Historical earthquake distribution map of the western Sichuan region. The inset map at lower left shows the study area. Red dots are earthquakes with magnitude 5.0 or greater from 780 BC to 2007 (China Earthquake Network Center catalog). The red star marks the epicenter of the M_S 8.0 Wenchuan earthquake of 12 May 2008 and its aftershocks to 12 June 2008 are represented by blue dots. Note: no $M > 7$ event had occurred in the middle segment of the North-South seismic belt (i.e. Longmenshan region) at least in the last 2000 years before 2008 (from Meng et al., 2008).

086 | *The Wenchuan Earthquake of 2008*

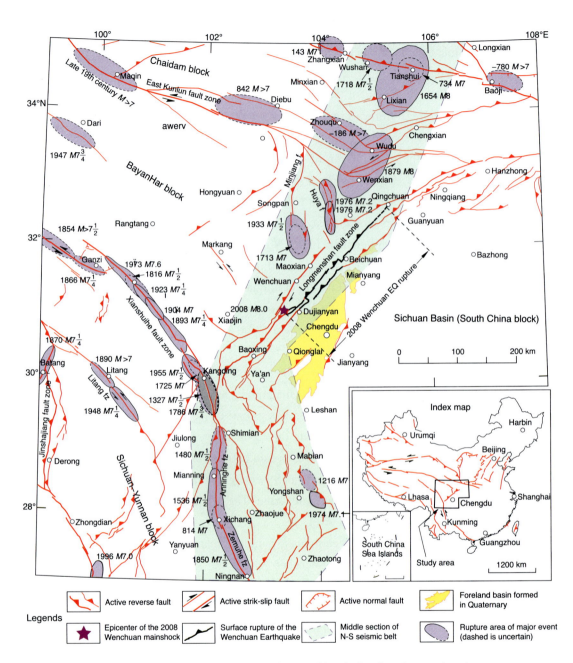

Figure 2.12 A map of the "source areas" of $M > 7$ historical and modern earthquakes on the middle segment of the North-South seismic belt. There are three main faults, forming a "Y" shape. Northwest: Xianshuihe fault; Northeast: Longmenshan fault; North-South: Anninghe fault (from Wen et al., 2009b).

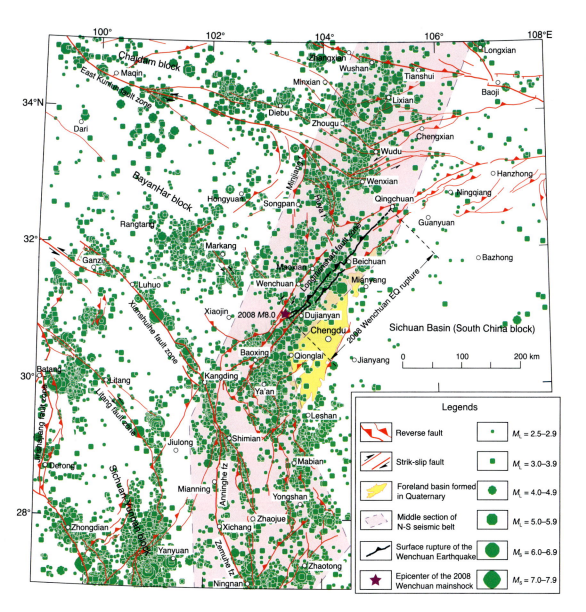

Figure 2.13 Seismicity on the middle segment of the North-South seismic belt and nearby in the past 20 years before the 2008 Wenchuan M8.0 earthquake, from Chinese seismic network $M > 2.5$ earthquake data from 12 May 1988 to 11 May 2008 (from Wen et al., 2009b).

Aftershocks

The Wenchuan earthquake had a large number of aftershocks. Dozens of $M > 5$, hundreds of $M > 4$, and thousands of $M > 3$ aftershocks occurred, which were well recorded by permanent and portable seismic stations. Sichuan seismograph network had compiled a list of 21,000 aftershocks by the end of 6 August 2008 (87 days after the mainshock).

Figure 2.14 shows the Magnitude–time (M–t)plot and the daily frequency of the Wenchuan aftershocks from 12 May to 6 August 2008.

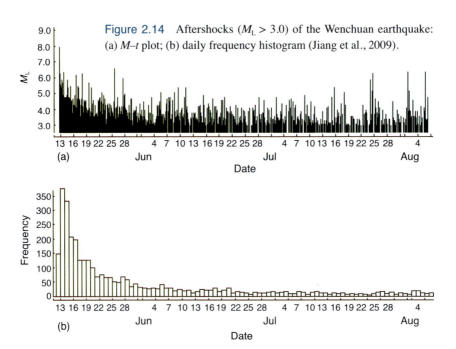

Figure 2.14 Aftershocks ($M_L > 3.0$) of the Wenchuan earthquake: (a) M–t plot; (b) daily frequency histogram (Jiang et al., 2009).

It can be seen that the daily frequency of aftershocks decayed quickly, but large aftershocks with magnitude greater than 5 occurred even 80 days after the mainshock.

Huang et al. (2008) selected 2706 $M > 2.0$ aftershocks, which were recorded by at least 8 stations, for relocation along with the mainshocks (Figure 2.15). The aftershock epicenters of the Wenchuan earthquake were distributed in the NE–SW direction, over a total length of about 330 km, which is about 100 km longer than the surface geological survey. It appears that substantially more aftershocks occur in regions with crystalline bedrock (Zheng et al., 2009).

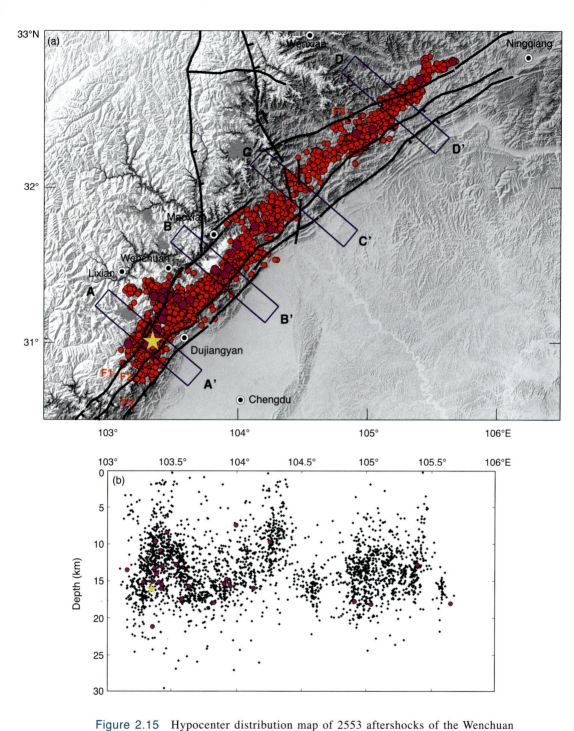

Figure 2.15 Hypocenter distribution map of 2553 aftershocks of the Wenchuan earthquake (modified from Huang et al., 2008).
(a) Epicenter distribution: mainshock (star), aftershocks with magnitude greater than 2.0 (red circles); 33 aftershocks with $M > 5.0$ (solid pink circles). Three faults of the Longmenshan fault zone: F1—back-range fault, F2—central fault and F3—front-range fault. Rectangles A, B, C and D are the locations for depth profiles which are shown in Figure 2.17. (b) Focal depth profiles along the longitude axis.

The Longmenshan fault zone consists of the back-range fault (F1), the central fault (F2), and the front-range fault (F3) from west to east. The aftershocks were concentrated on the west side of the central fault L2 of the Longmenshan fault zone.

The depth of the mainshock is 16 km, while focal depths of the aftershocks are predominantly between 5 km and 20 km (average depth 13.3 km, Figure 2.16).

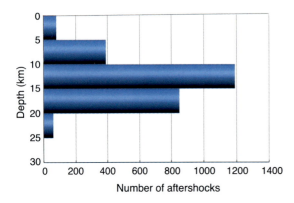

Figure 2.16   Number of aftershocks versus focal depth (2707 events) (redrawn from Huang et al., 2008).

The depth profiles reveal that the focal depth distribution in some areas is characterized by a high-angle westward dip (Figure 2.17). The rupture mode of the main shock shows reverse faulting in the south, with a large strike-slip component in the north.

Zheng et al. (2009) obtained focal mechanisms of the large aftershocks ($M_S \geqslant 5.6$) in Table 2.3. While most of those aftershocks show a thrust mechanism, there are some strike-slip earthquakes at the northernmost end of the rupture (Figure 2.18).

The focal mechanisms show that events located on the southern part of the central Beichuan-Yingxiu fault (BY) are mainly thrust earthquakes, which is consistent with the initial mechanism of the mainshock rupture. On the northern part, the aftershocks along BY are dominated by strike slip, which is quite different from the thrust slip rupture of the mainshock. In addition, the focal mechanisms have changed significantly from May 2008 to July 2009. The focal mechanisms of events 9 and 10 are quite different from those of the other earthquakes in the northern region. This suggests that the seismogenic stress field may change with time, and some new faults created or unknown faults reactivated.

Most great earthquakes are followed by numerous aftershocks. What time interval occurs between the mainshock and largest aftershock? Yu Yanxiang studied 44 global earthquakes, and the resulting statistics showed

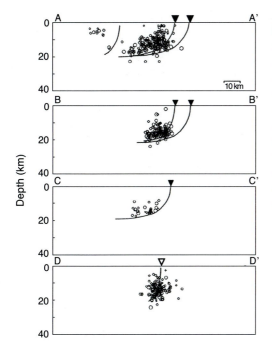

Figure 2.17 Four focal depth profiles. Solid triangles in profiles A–A', B–B' and C–C' (of Figure 2.15) represent intersections of faults with the surface, as confirmed by field reconnaissance after the earthquake; the open triangle in profile D–D' (of Figure 2.15) represents an inferred intersection. The fault traces are those inferred by Huang et al. (from Huang et al., 2008).

Table 2.3 Catalog of aftershocks ($M_s \geqslant 5.6$)

ID	Latitude (°N)	Longitude (°E)	Occurrence time	Magnitude
1	31.4	103.6	2008-05-12 19:10:58.4	6.0
2	31.4	104.0	2008-05-13 4:08:50.1	5.7
3	30.9	103.4	2008-05-13 15:07:01.0	6.1
4	31.3	103.4	2008-05-14 10:54:36.5	5.6
5	31.4	103.2	2008-05-16 13:25:49.0	5.9
6	32.23	105.0	2008-05-18 1:08:23.4	6.0
7	32.6	105.4	2008-05-25 16:21:46.9	6.4
8	32.8	105.6	2008-05-27 16:37:53.1	5.7
9	32.8	105.5	2008-07-24 3:54:46.5	5.6
10	32.8	105.5	2008-07-24 15:09:28.6	6.0
*	31.0	103.5	2008-05-12 14:43:15.0	6.0

Event ID * is an aftershock which happened only 15 minutes after the main shock, and is neglected by Zheng et al. (2009).

that most of the largest aftershocks occurred within 40 days of the mainshock (Figure 2.19). The largest Wenchuan earthquake aftershock occurred on 25 May 2008 ($M = 6.4$), 13 days after the main shock.

On 18 May, six days after the Wenchuan mainshock, CEA predicted an aftershock of $M6-7$ to occur during 19–20 May, and the Sichuan provincial government issued a warning using running text on television. This was a false alarm that led to confusion and embarrassment at the worst time. The prediction of strong aftershocks is still a difficult scientific problem.

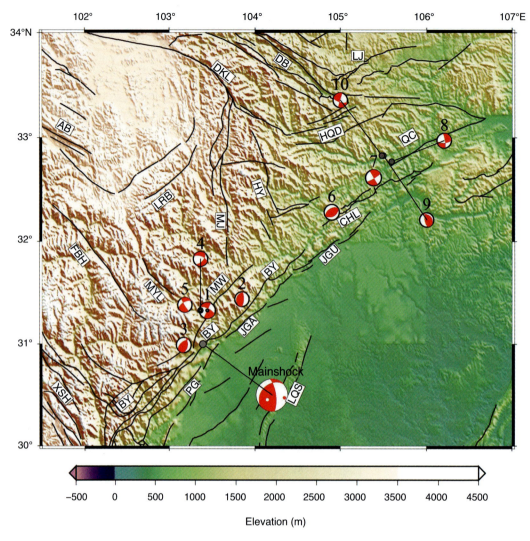

Figure 2.18 Focal mechanisms of Wenchuan earthquake aftershocks with $M_S \geqslant 5.6$. The numbers 1–10 correspond to the aftershock index numbers in Table 2.3. The focal mechanism shown for the Wenchuan mainshock is the Harvard CMT solution. The black lines represent Quaternary active faults (from Zheng et al., 2009).

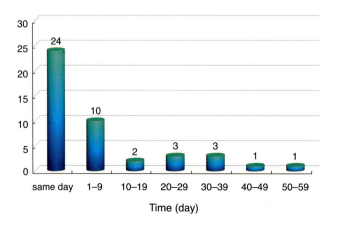

Figure 2.19 The time intervals between a mainshock and the largest aftershock (statistics of 44 global earthquake sequences) (Li Ming and Yu Yanxiang, personal communication).

2.3 Geophysical investigations before the earthquake

Geophysical investigations began with the study of seismicity.

North-South seismic belt

Figure 2.20 shows the epicenter distribution of earthquakes of $M \geqslant 6$ in China since 2000 BC. Each earthquake is shown by a circle whose diameter L is the following function of magnitude M:

$$\log L = \frac{M}{2} - 1.94$$

From Figure 2.20 it can be seen that the distribution is neither uniform or random, nor possesses a regular geometry. Most epicenters show spatial clustering in many different seismicity belts, and the remaining few epicenters show a random distribution. Because the distribution so complex and irregular, there are many methods for defining seismic belts in China. The North-South seismic belt is the most important belt (dotted line in Figure 2.20): this limited area contains almost a quarter of the earthquakes in mainland China.

More exactly, the North-South seismic belt is tilted North-Northeast with a range in latitude of 22°–38°N and in longitude of 97°–107°E. The southern part of the North-South seismic belt coincides with the Hengduan Mountain Range, but its northern part extends further northward than the Hengduan Mountains.

Forty earthquakes occurred in mainland China since 1900 with magnitude greater than 7.5, of which ten occurred in this belt, so the earthquakes in the North-South seismic belt are about a quarter of the total number in China (Table 2.4).

The Hengduan Mountain Range is a well-known geographical feature. It is a large mountainous region in China (Latitude: 22°–32°N, Longitude: 97°–103°E), forming the south-eastern part of the Tibetan Plateau, and adjacent to the west of the Sichuan Basin. The mountainous region occupies most of the western part of the present-day Sichuan Province, as well as the northwestern corner of Yunnan Province and the easternmost section of Tibet Autonomous Region. Mountain ridges at the southern end of the Hengduan Mountains form the border between Burma and China.

094 | The Wenchuan Earthquake of 2008

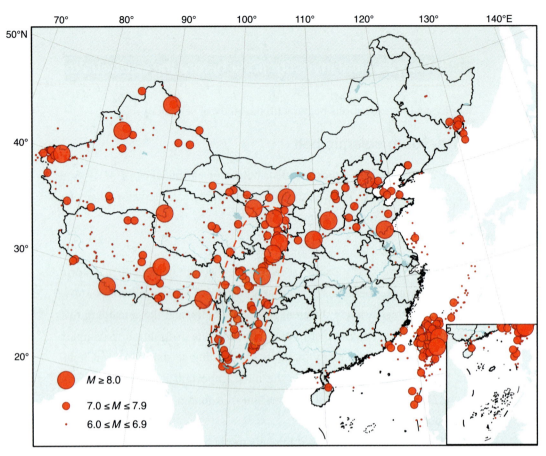

Figure 2.20 Epicenter distribution of major earthquakes ($M \geqslant 6$) of China from 789 BC to 14 Oct 2008. Red dotted line indicates North-South seismic belt; blue dotted line indicates the Hengduan Mountains.

Table 2.4 Earthquakes ($M > 7.5$) in North-South seismic belt since 1900

Date	Latitude (°N)	Longitude (°E)	M	Location
1920.12.16	36.42	104.54	8.5	Haiyuan, Ningxia
1927.05.23	37.42	102.12	8.0	Gulang, Gansu
1933.08.25	31.54	103.24	7.5	Diexi, Sichuan
1937.01.07	35.30	97.36	7.5	Alan Lake, Qinghai
1947.03.17	33.18	99.30	7.7	Dari, Qinghai
1955.04.14	30.00	101.48	7.5	Kangding, Sichuan
1970.01.05	24.12	102.36	7.8	Tonghai, Yunnan
1973.02.06	31.18	100.42	7.6	Luhuo, Sichuan
1988.11.06	22.30	100.00	7.6	Lanchang, Yunnan
2008.05.12	28.30	103.24	8.0	Wenchuan, Sichuan

The Hengduan Mountains consist of many mountain ranges, most of which run roughly north to south. Among them are the great Daxue mountain and Qionglai ranges, defining the eastern edge of the Tibetan Plateau and the western rim of the Sichuan Basin. In the southwestern part of the Hengduan Mountains region, three great rivers of China and Southeast Asia—the Yangtze, Mekong and Salween, run in deep parallel valleys separated by mountain ranges that are components of the Hengduan system. The most obvious feature of the Hengduan Mountains region is the combination of high mountains and deep valleys, where the relative elevation difference may reach 2000–3000 m over short distances, so that east-west traffic is very inconvenient, if not cut off altogether.

The tectonic activity which generates the Hengduan Mountain system is described later in section 2.4, from which it will be seen that the Wenchuan earthquake was to be expected on the Longmen Mountain Fault System (LMFS) of the North-South seismic belt, and is not surprising.

Bouguer gravity anomalies

Gravity is the force that causes two particles to pull towards each other. Gravity surveys can provide information on mass distribution beneath the surface of the Earth, as well as information on ongoing geological processes.

In geophysics, the usual gravity model is described with reference to the surface of a global spheroid (Hayford's reference ellipsoid, or WGS84) by rather simple formulae (2nd order functions of latitude). In order to determine the nature of the gravity anomaly due to the subsurface, a number of corrections must be made to the measured gravity value:

1. The theoretical gravity value should be removed, in order to show only local effects.
2. The elevation of the point where each gravity measurement was taken must be reduced to a reference datum to allow comparison of the whole profile. This is called the free-air correction, and when combined with the removal of theoretical gravity provides the free-air anomaly.
3. The normal gradient of gravity (change of gravity for change of elevation), in free air, is usually 0.3086 mGal/m. The Bouguer gradient of 0.1967 mGal/m (19.67 μm/(s^2·m)) assumes a mean rock density (2.67 g/cm^3) beneath the point of observation; this value

is found by subtracting the gravity due to the rock layer above the reference spheroid, which is 0.1119 mGal/m (11.19 μm/(s²·m)) for this density. Basically, we correct for the effects of any material between the point where gravimetry is done and the geoid. To do this, we model this material as being made up of a number of slabs of thickness t. These slabs have no lateral variation in density, but each slab may have a different density than the one above or below it. This is called the Bouguer correction; when applied to the free-air correction it provides the Bouguer anomaly

Bouguer anomalies are usually negative in mountain ranges because of isostasy, as the rock density of mountain roots is lower, compared with the surrounding Earth's mantle. Typical anomalies in the Central Alps are −150 mGal (−1.5 mm/s²). Very local anomalies are used in applied geophysics: if they are positive, they are possible indicators of metallic ores. At regional scales between entire mountain ranges and local ore bodies, Bouguer anomalies may indicate rock types and geological processes.

Bouguer gravity anomalies in the Tibetan Plateau (Figure 2.21) show a zone of low values (about −550 mGal (−5.5 mm/s²)), gradually increasing to the north and east. The general trend in western China is: east high and west low; north high and south low. In addition, there are several small and different patterns of local anomalies with steeper variations. With these superimposed on the general trend, this forms a complex Bouguer gravity anomaly field.

The Wenchuan earthquake is located in the highest Bouguer gravity anomaly gradient zone, and in a highly seismic zone. The equilibrium state of the crust depends on gravity and the role of tectonic forces: gravity moves the crust towards equilibrium, whereas the interplay of tectonic forces usually destroys the state of crustal equilibrium.

Gravity and geodetic data show that the Tibetan Plateau is far from isostatic equilibrium, and large scale uplift is the main driving force for continent dynamics. The North-South seismic belt (particularly the Longmenshan fault zone) is located in the highest gradient of Bouguer gravity anomaly, thus there is great seismic hazard.

Crustal thickness

Figure 2.22 shows the distribution of crustal thickness in China. The thickest

Chapter 2 · Seismological features | 097

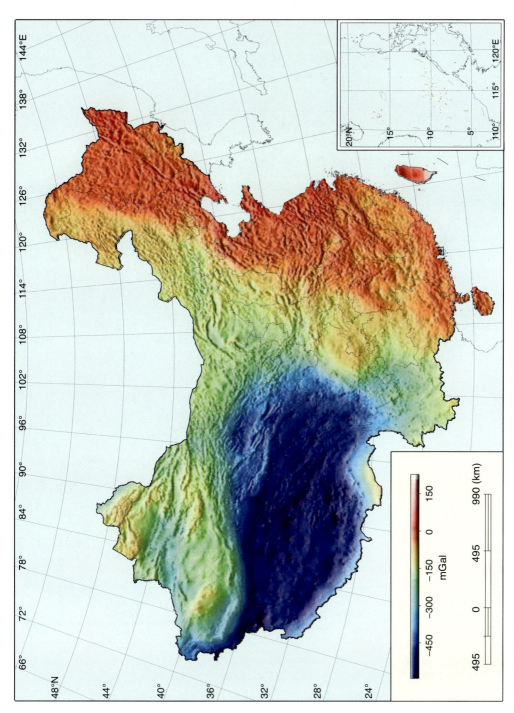

Figure 2.21 Bouguer gravity anomalies map (2011) of China (courtesy of Li Jiancheng, Wuhan University).

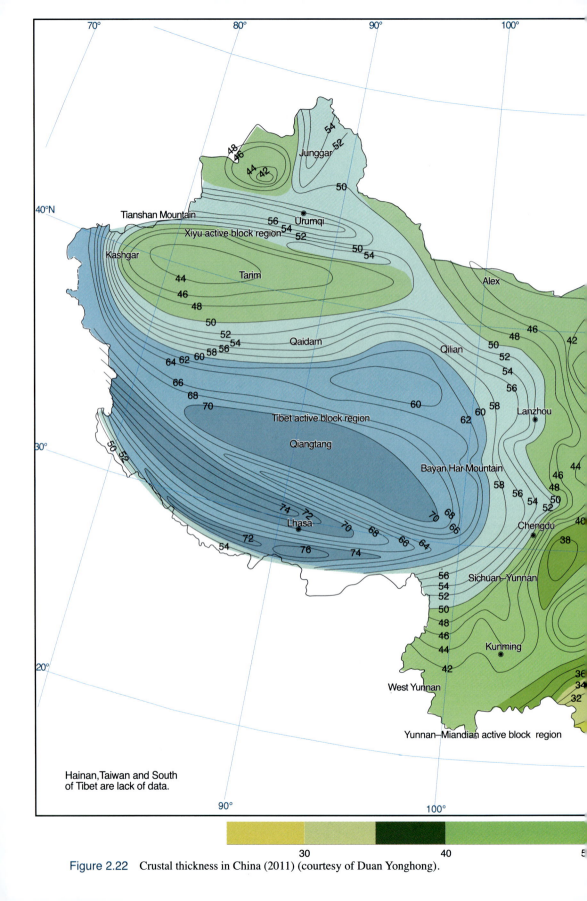

Figure 2.22 Crustal thickness in China (2011) (courtesy of Duan Yonghong).

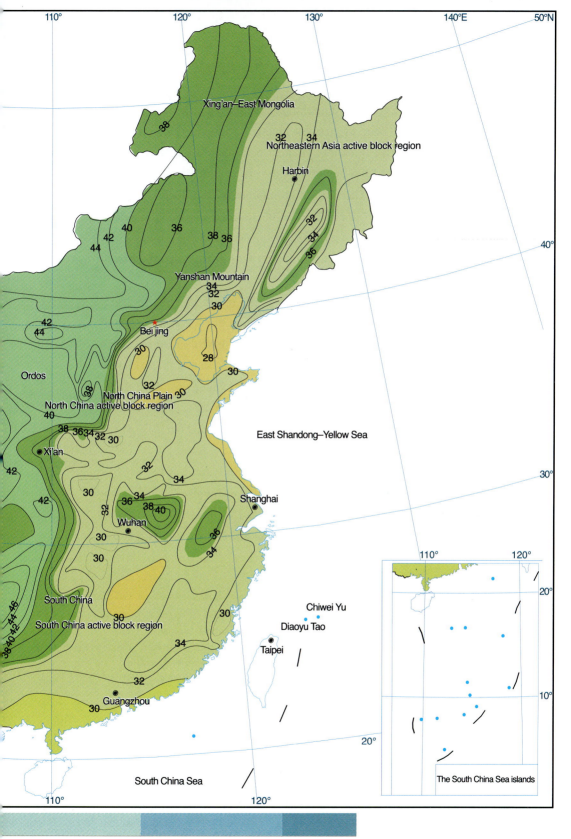

crust (60–70 km) is under Tibet, from which, crust thickness decreases eastwards to 30–40 km. The Wenchuan earthquake in Longmenshan region occurred in the zone of steepest change in crustal thickness. The high topography and crust thickness gradient belt, termed the North-South seismic belt, crosses the LMFS and divides the crust of mainland China into eastern and western parts. There are high mountains and deep valleys in the LMFS. Within 50 km from east to west, the average elevation increases sharply from 2.5 km to 5.5 km, with the highest peak at 7.5 km (Figure 2.23). Towards the centre of the LMFS, the thickness of the crust increases sharply from 30 km to 70 km.

Crustal thickness provides a record of the evolution of the crust. Consisting mostly of granitic rock, continental crust has a density of about 2.7 g/cm^3 and is less dense than the material of the Earth's mantle, which consists of mafic rock. Continental crust is also less dense than oceanic crust, though it is considerably thicker; mostly 25–70 km against the average oceanic crustal thickness of around 7–10 km. About 40% of the Earth's surface is underlain by continental crust, which makes up about 70% of the volume of Earth's crust.

There are currently about 7 billion km³ of continental crust, but this quantity varies due to the nature of the tectonic forces involved. The relative permanence of continental crust contrasts with the short life of oceanic crust. Since continental crust is less dense than oceanic crust, when active

Figure 2.23 Rugged topography of the Sichuan region of China. This view southwest shows rapids through a landslide deposit in the river gorge, with the snow-capped mountains of the Gonga Massif in the distance. Here, the Longmenshan River is at an elevation of ~ 1100 m. Mountain peaks to the west rise to elevations greater than 7000 m over less than 30 km, making this region one of the most dramatic examples of topographic relief on Earth (from Kirby et al., 2008).

margins of the two meet in subduction zones, the oceanic crust is typically subducted into the mantle. Continental crust is only rarely subducted; it may occur where continental crustal blocks collide and overthicken, causing deep melting under mountain belts such as the Himalayas or the Alps. Thus continental crust and the rock layers that lie within it are the best archive of Earth history.

The height of mountain ranges is usually related to the underlying thickness of crust. This results from the isostasy associated with orogeny (mountain formation). The crust is thickened by compressive forces related to subduction or continental collision. The buoyancy of the crust forces it upwards, then the forces of uplift are balanced by gravity and erosion. This forms a keel or mountain root beneath the mountain range, where the thickest crust is found. The thinnest continental crust is found in rift zones, where the crust is thinned by detachment faulting and eventually severs, being replaced by oceanic crust.

Today continental crust is produced and (far less often) destroyed mostly by plate tectonic processes, especially at convergent plate boundaries. Additionally, continental crustal material is transferred to the oceanic crust by erosion and sedimentation. New material can be added to the continents by the partial melting of oceanic crust at subduction zones, causing the lighter material to rise as magma, forming volcanoes. Also, material can be accreted "horizontally" when volcanic island arcs, seamounts or similar structures collide with the side of the continent as a result of plate tectonic movements. Continental crust is lost, due to erosion and sediment subduction, tectonic erosion of forearcs, delamination, and deep subduction of continental crust in collision zones. Many aspects of crustal growth are controversial, such as rates of crustal growth and recycling, whether the lower crust is recycled differently than the upper crust, over how much of Earth's history plate tectonics has operated, and if it has been the dominant mode of continental crust formation and destruction. It is a matter of debate whether the amount of continental crust has been increasing, decreasing, or constant over geological time. One model indicates that the growth of continental crust appears to have occurred in spurts of increased activity corresponding to five episodes of increased production through geologic time (Taylor et al., 1995).

Low-velocity zones

Low-velocity zones (LVZ) are important features in geophysics. The existence of a subcrustal low-velocity zone was first proposed from the observation of slower than expected seismic wave arrivals from earthquakes in 1959 by Beno Gutenberg. He noted that between 1° to 15° from the epicenter the longitudinal arrivals showed an exponential decrease in amplitude after which they showed a sudden large increase.

The subcrustal low-velocity zone occurs close to the boundary between the lithosphere and the asthenosphere in the upper mantle; it is called the Lehmann discontinuity and occurs at about 220±30 km depth. It is characterized by unusually low seismic shear-wave velocity compared to the surrounding depth intervals. This range of depths also corresponds to anomalously high electrical conductivity. A second low-velocity zone has been detected in a thin ~ 50 km layer at the core-mantle boundary. These LVZs may have important implications for plate tectonics and the origin of the Earth's crust.

The LVZ has been interpreted to indicate the presence of a significant degree of partial melting, where a very limited amount of melt (about 1%) is needed to produce these effects. Water in this layer can lower the melting point, and may play an important part in its composition.

Wei et al. (2010) used 71,670 P-wave arrival times from 3594 earthquakes recorded by the Sichuan and Yunnan seismic networks to determine the three-dimensional P-wave velocity structure in the crust and uppermost mantle beneath the southeastern Tibetan Plateau (Figure 2.24).

Their results show that prominent low P-wave velocity (low-V_P) anomalies exist in the mid to lower crust west of Longmenshan. In contrast, a high P-wave velocity (high-V_P) anomaly is detected in the middle and lower crust beneath the Sichuan Basin. This result provides seismic evidence for a dynamic model of lower crustal flow. Ongoing lower crustal flow beneath the central and eastern Tibetan Plateau presses against the mechanically strong Sichuan Basin, resulting in accumulated strain in the Longmenshan region.

Lei and Zhao (2009) found that the source area of the Wenchuan mainshock is underlain by obvious low-velocity anomalies, suggesting that fluids may exist within the fault zone (Figure 2.25). These fluids may affect the occurrence of the large Wenchuan earthquake. The P-wave tomography provides significant seismological evidence for the upward intrusion of the lower crustal flow along the Longmenshan fault zone.

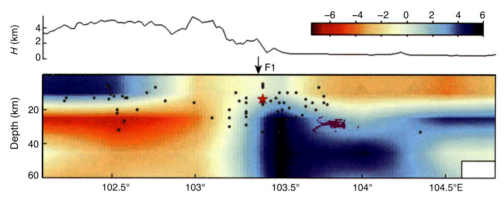

Figure 2.24 Vertical cross sections of P-wave tomography along the line AB, where AB is a line perpendicular to the Longmenshan fault through Wenchuan. Surface topography is shown above the profile. Red and blue colors represent slow and fast velocity perturbations (%), respectively, with the scale shown at the bottom. Grey dots indicate earthquakes with magnitude 4.0 that occurred within a 15 km wide band along the profile. The red star denotes the Wenchuan mainshock. Note that most of the large crustal earthquakes ($M > 4.0$) in this region are located in transitional areas between low-V_P and high-V_P zones (from Wei et al., 2010).

Figure 2.25 P-wave tomographic images along the east-west direction (perpendicular to the trend of fault) passing through the source area of the Wenchuan mainshock. The location of the profile is shown in the map below. The star denotes the Wenchuan mainshock; blue colors denote high velocity; red colors denote low velocity. The uppermost curve is the elevation along the profile (from Lei et al., 2009).

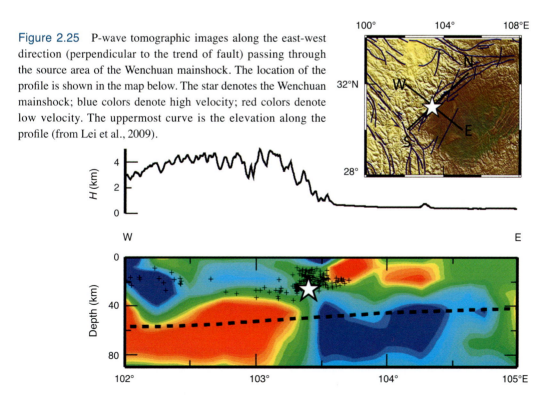

2.4 Tectonic setting of Wenchuan earthquake

Seismotectonics is the study of the relationship between earthquakes, active tectonics and individual faults of a region. It provides an understanding of which faults are responsible for seismic activity in an area by analysing a combination of regional tectonics, recent instrumentally recorded events, accounts of historical earthquakes and geomorphological evidence. This information can then be used to quantify the seismic hazard of an area.

The Wenchuan earthquake occurred within a background of long-term uplift and eastward enlargement of the Tibetan Plateau. Geological, geodetic, and geophysical data allow us to place this major seismic event in context with respect to ongoing deformation along the eastern margin of the Plateau.

The earthquake occurred as a result of motion on a northeast striking thrust fault that runs along the margin of the Sichuan Basin. The seismicity of Central and eastern Asia is caused by the northward movement of the India plate at a rate of 5 cm/year and its collision with Eurasia, resulting in the uplift of the Himalaya and Tibetan Plateau and associated earthquake activity. This deformation also results in the extrusion of crustal material from the high Tibetan Plateau in the west towards the Sichuan Basin and southeastern China. The eastern margin of Tibetan Plateau frequently suffers large and deadly earthquakes. In August 1933, the magnitude 7.5 Diexi earthquake, about 90 km northeast of today's earthquake, destroyed the town of Diexi and surrounding villages, and caused many landslides, some of which dammed rivers.

Tibetan Plateau

The Tibetan Plateau (25°–40°N, 74°–104°E) is a vast, elevated plateau in Central Asia. It occupies an area of around 1000 km by 2500 km, and has an average elevation of over 4500 m. Sometimes called "the roof of the world", or "the third pole" (after the Antarctic and Arctic poles) . It is the highest, largest and youngest Plateau, with an area of 2.5 million km^2 (Figure 2.26).

The Tibetan Plateau is bordered by large mountain ranges. It is bordered to the north by the Kunlun Range which separates it from the

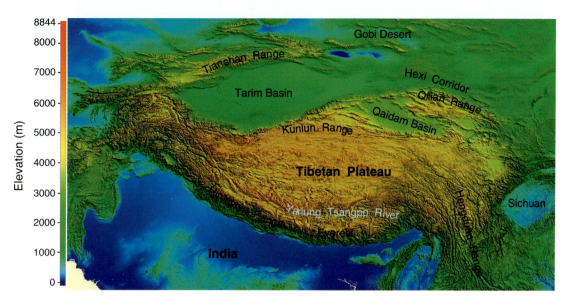

Figure 2.26　The Tibetan Plateau and surrounding areas.

Tarim Basin, and to the northeast by the Qilian Range which separates the plateau from the Hexi Corridor and Gobi desert. Near the south the plateau is transected by the Yarlung Tsangpo River valley which flows along the base of the Himalayas. To the east and southeast the Plateau gives way to the forested gorge and ridge geography of the mountainous headwaters of the Yangtze (Jinsha) , Lancang (Mekong) and Nujiang (Salween) rivers in western Sichuan (the Hengduan Mountains) and southwest Qinghai. In the west it is bounded by the curve of the rugged Karakoram Range of northern Kashmir.

　　It is generally believed that no continental crust originally existed on the Earth. All continental crust ultimately derives from the fractional differentiation of oceanic crust, a process which has been active over many eons and continues today.

　　There is conclusive evidence that the geological history of the Tibetan Plateau can be traced back 400–500 million years (the Ordovician). Up to 280 million years ago (Early Permian), the vast Tibetan Plateau was a large ocean, which connected with other oceans in North Africa, Southern Europe, West Asia and southeast Asia areas, called the Tethys Ocean (or "ancient Mediterranean").

　　The Tethys Ocean had a warm climate, with abundant marine animal and plant development. The north side and south sides of the Tethys were the split parts of an earlier continent, known as Pangea. The south side, called Gondwana, included the present South America, Africa, Australia, Antarctica and South Asian sub-continents. The north side of the Tethys was called Eurasia, and also known as Laurasia. It included the present Europe, Asia and North America.

The Tibetan Plateau is the world's youngest Plateau, which is generated by plate motion beginning 240 million years ago, when the separated Indian plate moved northward at a rapid velocity, and the Tethys ocean started to close in the north–south direction. First, the northern region of the present Tibetan Plateau (Kunlun Mountains and Hoh Xil) was uplifted and ocean floor became land. Later, about 210 million years ago, the central region (Qiangtang region, Karakoram Mountains, Tanggula) of the present Tibetan Plateau emerged out of the sea and became land. Then from 80 million years ago, the continuing northward drift of Indian plate, once again caused a strong tectonic movement. The whole present Tibetan Plateau rose from the ocean to become land. The Plateau landscape shape formed, which is called the Himalayan tectonic movement in Geology. The Tibetan Plateau uplift process is not uniform in this process; it is not a single surge, but a rise through several different stages.

The Tibetan Plateau is the result of the post-collision north–south convergence of India and Eurasia from approximately 50 Ma. The uplift of the Tibetan Plateau has been accompanied by both complex tectonic movements and deep dynamic processes. In contrast to the flat central Plateau, most of the marginal mountain belts bordering the Plateau exhibit steep topographic gradients. To interpret the crustal thickening and surface uplifting without significant late Cenozoic crustal shortening along the eastern margin of the Plateau, a dynamic model of the lower crustal flow was needed.

From about 10,000 years ago, the Plateau uplift has become faster, at an average annual rate of 7 cm/year, making the Plateau today's "Roof of the World".

Under the rapidly uplifting Tibetan Plateau, there is an S-wave LVZ which covers a large area. Figure 2.27 shows the S-wave velocity at 40km depth, and it is clear that China's largest S-wave LVZ lies underneath the Tibetan Plateau.

Today, weathering is the main process in the central part of the Tibetan Plateau, and uplift is the dominant process on the boundary margins of the Plateau. The long-term uplift and eastward enlargement of the Tibetan Plateau are the main cause of the Wenchuan earthquake.

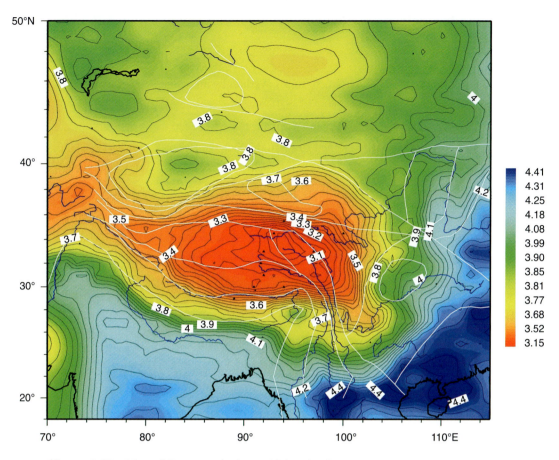

Figure 2.27 Map of S-wave velocity at 40 km depth under the Tibetan Plateau and surrounding regions (from Shu et al., 2002).

Longmenshan

Geological surveying of the Longmenshan fault zone began in the 1920s. The Chinese geologist Zhao Yazeng first discovered "klippe" structure, typical of thrust fault terranes, within the Longmenshan fault zone (at Pengzhou) in the late 1920s. Studies by other geologists confirmed the feature as a text-book example of a typical thrust. Since the 1980s, with the application of the theory of plate tectonics, the study of the Longmenshan fault zone has been more detailed (Figure 2.28).

The Cenozoic deformation of the Longmenshan is superimposed on a preexisting Mesozoic orogeny. This older deformation provided the starting geometry for later Cenozoic deformation (see Burchfiel et al., 2008, Figure 2.29). Mesozoic deformation in the Longmenshan took place in Late Triassic and Jurassic time, when two distinct structural sequences were deformed and juxtaposed by thrust faulting. The autochthonous lower sequence consists mainly of Late Precambrian basement rocks overlain by an incomplete section of Late Proterozoic to Middle Triassic shallow-water sedimentary rocks and Upper Triassic–Jurassic clastic rocks that appear to be foredeep basin deposits and grade eastward into finer-grained strata in the Sichuan Basin.

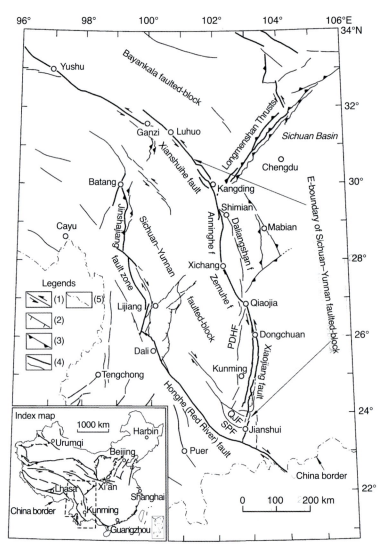

Figure 2.28 A simplified map of active faults in and surrounding the Sichuan–Yunnan faulted-block of southwestern China. Names of major faults and faulted-blocks are labeled. The inset map shows the position of the Sichuan–Yunnan faulted-block in continental China. Legend notation: (1) active strike-slip fault; (2) active normal fault; (3) active reverse fault or thrust; (4) major (thick line) and secondary (thin line) active faults; (5) political border. Abbreviations for secondary fault names: QJF, Qujiang fault; SPF, Shiping fault; PDHF, Puduhe fault (from Wen et al., 2009a).

The Longmenshan marks not only the present boundary between the high topography of the Tibetan Plateau to the west and the relatively undeformed Sichuan Basin to the east, but also the limit of deformation during the Mesozoic Indosinian orogeny. During the Late Triassic to Early Jurassic, a sequence of continental margin sediments and flysch deposits were highly deformed and thrust eastward onto the rocks of the Yangtze craton while the Sichuan Basin was accumulating clastic sediments as a foredeep basin.

The Sichuan Basin is roughly circular, containing primarily Mesozoic and Paleozoic sedimentary rocks more than 10 km deep, and rimmed along its southern margin by Cenozoic structures that merge westward and northward into the Longmenshan. To the north and east, the surrounding ranges are folded belts of Late Triassic–Cretaceous and Late Cretaceous age, respectively. Thus, the basement beneath the basin remained relatively undeformed during the Mesozoic and Cenozoic deformations that affected the surrounding regions.

The Longmenshan and the Minshan are asymmetric ranges bounded by steep, high-relief margins on their eastern sides and only modest western slopes. To the west, elevations rise toward > 5000 m on the Tibetan Plateau. The steepest margin on the eastern Tibetan Plateau occurs where the Longmenshan borders the Sichuan Basin; elsewhere, the Tibetan Plateau margin is gently sloping. To the north, the crest of the Longmenshan deviates westward from the range front and continues into the Minshan (Figures 2.29, 2.30). Much of the active convergence along the east side of the Tibetan Plateau follows the high topography of these ranges.

In the vicinity of the Wenchuan earthquake, the eastern margin of the Tibetan Plateau rises steeply westward from an elevation of 500 m to over 4000 m (Figure 2.30). Mountain peaks within the Longmenshan reach elevations higher than 6000 m. The eastern Tibetan Plateau margin formed by the Longmenshan coincides with steep gradients in crustal thickness (from 60–65 km in the west to ~ 40 km in the east, Figure 2.22), and with steep gradients in gravity anomaly.

Seismically active faults and related fault-generated folds have a direct effect on the geomorphology of a region, and may allow the direct identification of active structures not previously known.

The stress field generating the Wenchuan earthquake and Longmen-

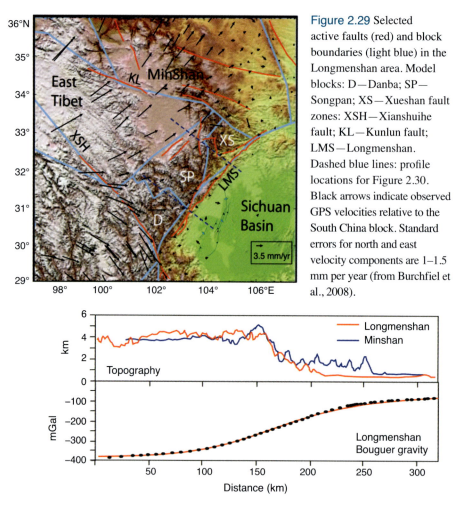

Figure 2.29 Selected active faults (red) and block boundaries (light blue) in the Longmenshan area. Model blocks: D—Danba; SP—Songpan; XS—Xueshan fault zones: XSH—Xianshuihe fault; KL—Kunlun fault; LMS—Longmenshan. Dashed blue lines: profile locations for Figure 2.30. Black arrows indicate observed GPS velocities relative to the South China block. Standard errors for north and east velocity components are 1–1.5 mm per year (from Burchfiel et al., 2008).

Figure 2.30 Topography profiles across the Longmenshan, and observed (dots) and computed (line) Bouguer gravity anomalies for Airy compensation of the Longmenshan for a density contrast between the crustal root and mantle of 400 kg/m^3. Profile locations are indicated in the Figure 2.29 by dashed purple lines (from Burchfiel et al., 2008).

shan's southeast push is caused by the collision of the Indian-Australian plate into the Eurasian plate and its northward push. This interplate motion has caused large scale structural deformation inside the Asian continent, resulting in thickening of the crust of the Tibetan Plateau, the uplift of its landscape and eastward extrusion. Near the Sichuan Basin, the Qinghai-Tibet Plateau's northeastward movement meets with strong resistance from the South China block, causing a high degree of stress accumulation in the Longmenshan thrust formation. This finally caused a sudden dislocation of the Yingxiu–Beichuan fracture, which generated the violent Wenchuan earthquake of $M_S 8.0$.

The Tibetan Plateau and Sichuan Basin are two very different geological units, separated by Longmenshan, where exist the largest variations of gravity, crustal thickness, and topography in the western margins of the Tibetan Plateau.

An intracontinental thrust fault

Earthquakes result from sudden motion on faults, which are usually dipping, with the upper part called the hanging wall, and the lower part the footwall. During an earthquake, the fault walls can move in three ways: (1) the two walls slip horizontally past each other (strike slip faulting); (2) the hanging wall moves upward and the footwall downward (thrust faulting); (3) the hanging wall moves downward and the footwall upward (normal faulting). A strike-slip fault is vertical or near-vertical and the rock on opposite sides of the fault moves horizontally. A thrust fault is generally a low-angle-fault in which rock on one side of the fault slides up and over rock on the other side (Figure 2.31).

Normal and thrust faulting are examples of dip-slip, where the displacement along the fault is in the direction of dip, and there is a vertical component of movement. Normal faults occur mainly in areas where the crust is being extended, such as a divergent plate boundary. Thrust faults occur in areas where the crust is being shortened, such as a convergent plate boundary. Many earthquakes are caused by movement on faults that have components of both dip-slip and strike-slip; this is known as oblique-slip.

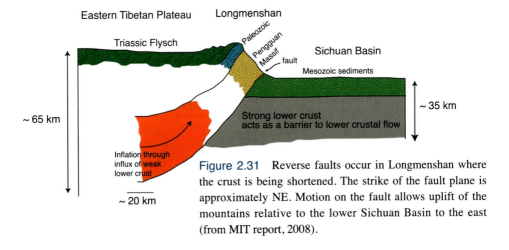

Figure 2.31   Reverse faults occur in Longmenshan where the crust is being shortened. The strike of the fault plane is approximately NE. Motion on the fault allows uplift of the mountains relative to the lower Sichuan Basin to the east (from MIT report, 2008).

The Longmenshan fault is a thrust fault which runs along the base of the Longmenshan Mountains, in Sichuan Province in southwestern China. The strike of the fault plane is approximately NE. Motion on this fault is responsible for the uplift of the mountains relative to the lowlands of the Sichuan Basin to the east.

From the discovery of the klippe structure in the Longmenshan area in the 1920s, to the thrust faulting of the 2008 Wenchuan earthquake, thrust fault movements have been shown to dominate, so that shortening or convergent characteristics were obvious. However, the area has numerous geological features not typical of active convergent mountain belts, including the presence of a steep mountain front (> 4 km relief) but an absence of large-magnitude low-angle thrust faults; young high topography (post ca. 15 Ma) and thickened crust but low shortening rates according to global positioning system (GPS) observations (< 3 mm/year); and no coeval foreland subsidence (Burchfiel et al., 2008).

The Wenchuan earthquake occurred in a thrust fault system, which is rare in continents. Thrust faults most commonly occur in oceanic subduction zones, where frequent oceanic interplate earthquakes occur; all large oceanic earthquakes show thrust fault movements. In contrast, most continental earthquakes have strike-slip fault mechanisms. However, most earthquakes in the margin of the Tibetan Plateau show thrust fault mechanisms (Figure 2.32).

The Wenchuan earthquake is very important, because it is rare to observe movement on a thrust fault inside a continent. It allows a better understanding of continental dynamics.

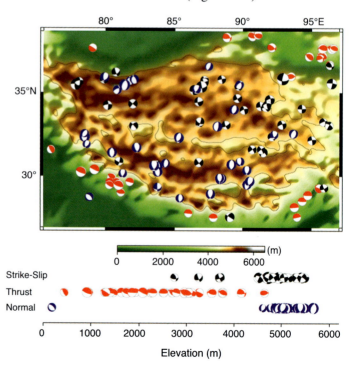

Figure 2.32 Focal mechanisms of earthquakes in the Tibetan Plateau and its surrounding region. Red: thrust faulting; blue: normal faulting; black: strike-slip faulting. Most earthquakes within the Plateau show normal faulting or strike-slip faulting, while earthquakes occurring at the boundaries of the Plateau show thrust faulting (from Elliott et al., 2010).

Weak crustal layer

The magnitude of Cenozoic shortening across the Longmenshan is variable but small, probably on the order of tens of kilometers.

In eastern Tibet, few Cenozoic shortening structures are observed in the field. The geometry of the Cenozoic shortening structures in the Longmenshan is such that only some of the dip-slip displacement on deep faults may reach the surface, while the rest may be absorbed by folding or flow within the overlying layers and in the Sichuan Basin (Figure 2.33).

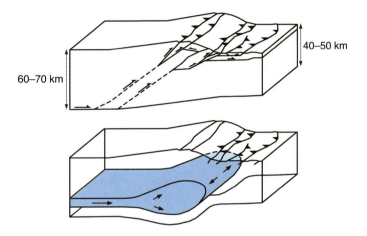

Figure 2.33 Alternative conceptual models for the uplift of the Tibetan Plateau and development of the Longmenshan range front. Top panel, uplift is produced by thrust faulting and crustal shortening. Bottom panel, uplift is produced by inflation of the ductile lower crust (from Hubbard et al., 2009).

One interesting point is that crustal thickening beneath the eastern Tibetan Plateau occurred without large-scale shortening of the upper crust but instead is caused by ductile thickening of the deep crust in a weak (low-viscosity) layer. Late Cenozoic shortening across the Longmenshan could be as little as 10–20 km, with folding and faulting mainly accommodating differential surface uplift between the Plateau and the Sichuan Basin. The 2008 Wenchuan earthquake is probably attributable to long-term uplift, with slow convergence and right-slip, of the eastern Plateau relative to the Sichuan Basin.

The duration of crustal thickening beneath the Longmenshan and eastern Plateau is probably similar to that of the surface uplift, which is younger than 15 Ma. Based partially on the absence of significant Late Cenozoic shortening structures south of the Kunlun fault, Clark and Royden (2000) proposed that crustal thickening in eastern Tibet occurred largely within a weak (low-viscosity) zone in the mid to lower crust. A variety of data are consistent with this interpretation (e.g. high crustal temperatures, slow seismic wave speeds, flat to gently dipping topographic surfaces). If this interpretation is generally correct, the zone of weak crust probably does not extend beneath the eastern Longmenshan, because crust with extensive zones of weakness at depth cannot support steep topographic gradients of significant lateral extent. Clark and Royden (2000) also postulated that the edge of the high Plateau may be narrowly localized along the Longmenshan because the mechanically strong lithosphere of the Sichuan Basin obstructs the eastward flow of weak crust at depth. In their concept, localization of Late Cenozoic deformation and active faulting along the Longmenshan is largely controlled by the rheological contrast between the weaker crust of eastern Tibet and the craton-like crust/lithosphere of the Sichuan Basin. They attributed the northeastward motion of the east Tibetan crust, relative to the Sichuan Basin, to the growth of the eastern Plateau to the northeast, with northeastward-moving crustal fragments within eastern Tibet diverted around the mechanically strong Sichuan Basin block.

The earthquake occurred as the result of motion on a northeast striking reverse fault (thrust fault) on the northwestern margin of the Sichuan Basin. The earthquake's epicenter and focal-mechanism are consistent with it having occurred as the result of movement on the Longmenshan fault or a tectonically related fault. The earthquake is a consequence of tectonic stresses resulting from the convergence of crustal material slowly moving from the high Tibetan Plateau, to the west, against strong crust underlying the Sichuan Basin and southeastern China.

In Figures 2.24 & 2.25, the main shock is located in the transition zone where seismic velocity changes dramatically, indicating that the crust near the source zone is extremely heterogeneous. It is also remarkable that a continuous low-V_P layer exists in the mid to lower crust beneath the southeastern Tibetan Plateau from the profile A–B (Figure 2.24). This suggests that the mid to lower crust is relatively mechanically weak there. Wei et al. (2010) proposed a lower crustal flow model to explain the cause of Wenchuan earthquake. Due to the northward subduction of the Indian plate, low-viscosity material in the lower crust beneath the central and eastern Tibetan Plateau has moved toward the east, where it is blocked by the mechanically strong Sichuan Basin. During this process, the strain accumulated slowly in the Longmenshan region. When the increasing strain energy exceeded a threshold level, the energy was suddenly released in the Wenchuan earthquake.

2.5　Seismic waves generated by earthquake

As the fault ruptured, it radiated seismic waves into the surrounding rock (Figure 2.34). The elastic waves propagated outwards, through the Earth's interior as P and S body waves, and along its surface as Love and Rayleigh surface waves. Surface waves are analogous to water waves and travel slower than body waves. Because of their low frequency, long duration, and large amplitude, they can be the most destructive type of seismic wave.

Seismic waves circle the Earth

The seismic wave generated by the Wenchuan earthquake travelled through the Earth, and the surface wave traveled round the Earth's surface more than once, as shown as Figure 2.35. The ground motion was dominated by surface waves (Rayleigh waves), which produced peak-to-peak amplitudes of at least 1 mm at even the most remote distances on the Earth's surface.

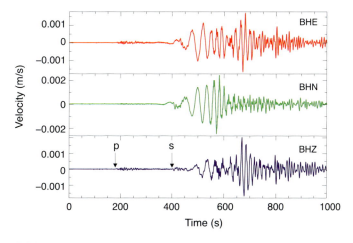

Figure2.34　Seismograph of Wenchuan earthquake recorded at Shenyang station.

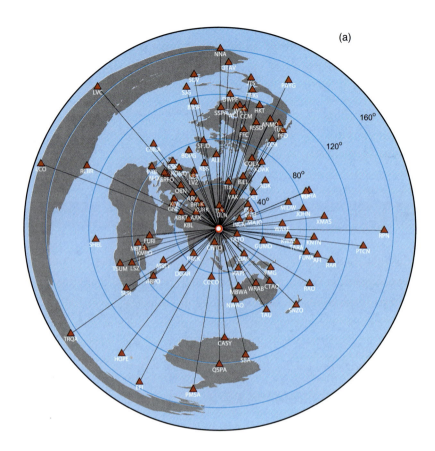

(a)

The energy of seismic waves released (E) is usually used to define the magnitude (M) of the earthquake, according to the formula

$$\log E = 11.8 + 1.5\,M$$

Thus the energy released by the Wenchuan earthquake calculated as $10^{23.8}$ erg, is equivalent to the detonation of 10,000,000 tons of TNT.

Ground motion felt at Beijing

The energy of an earthquake is released within seconds and spreads through the Earth as seismic waves. Wave velocity varies with the different wave types, and so does the damage caused at the surface. The P-wave is the fastest wave, but it usually causes less damage than the surface wave, which travels slower, makes the ground shake severely, and produces larger damage. After the mainshock on 12 May 2008, the P-wave and surface waves took 200s and 300s (or more) respectively to travel to Beijing, which is about 1530 km away from the epicenter.

Chapter 2 · Seismological features | 117

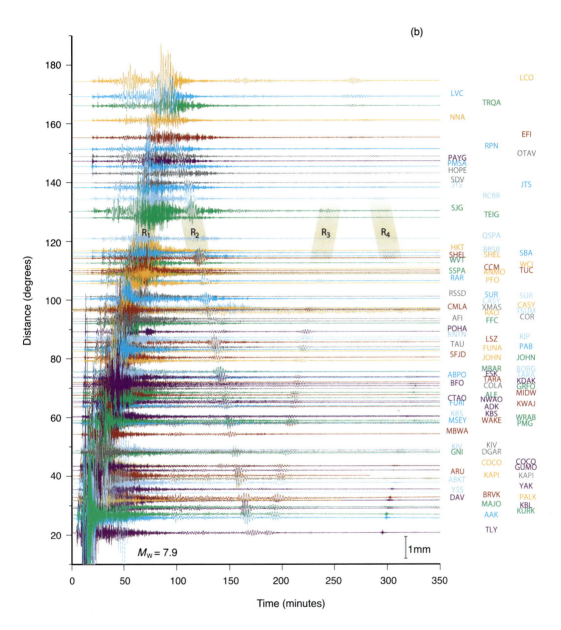

Figure 2.35 (a) Seismic waves of the Wenchuan earthquake recorded by seismographs over the globe. (b) Six hours of vertical ground shaking (at both GSN stations worldwide and NDSN stations in China) are displayed against distance from the source. The ground motions are dominated by surface waves (Rayleigh waves), which produced peak-to-peak amplitudes of at least 1 mm (see scale at lower right) all over the Earth's surface. The series of major arrivals at each station involve R1 (Rayleigh wave that travels along the minor great circle arc), R2 (Rayleigh wave traveling along the major great circle arc), R3 (the same pulse as R1, but with an additional global circuit), and R4 (the same pulse as R2 with an additional global circuit).

Figure 2.36 shows the seismic Raleigh waves recorded at the Beijing (BJT) earthquake observation station in Beijing. The maximum amplitude of ground motion is about 7 cm, and many inhabitants of high buildings in Beijing felt significant shaking 5 minutes after the Wenchuan earthquake.

Comparison with waves generated by Tangshan earthquake of 1976

The 2008 Wenchuan earthquake is a shallow intraplate earthquake, similar to the Tangshan earthquake (28 July 1976, $M_S 7.6$). However, in terms of strength, intensity and economic losses, the 2008 event was greater than the Tangshan earthquake. The ground rupture caused by the 2008 event is approximately 300 km, three times that of the Tangshan earthquake, and there was a greater duration time of co-seismic slip (120 sec).

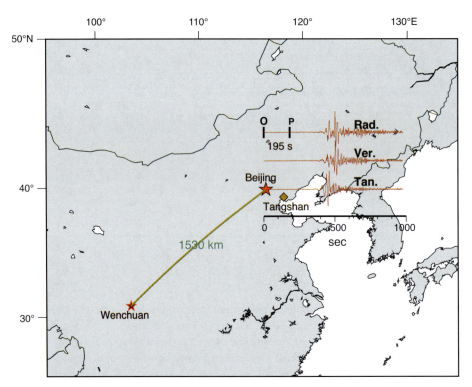

Figure 2.36 The amplitude of ground motion recorded at Beijing seismic station is 7 cm, with 5 sec wave period.

The Grafenberg (GRA1) array of Germany, which has a long observation history, recorded both the Wenchuan and Tangshan earthquakes. Figure 2.37 displays the waveforms of these two strong earthquakes recorded by GRA1. The distances from GRA1 to the epicenters of the Tangshan and Wenchuan earthquakes are around 7800 km and 7600 km, respectively. Since the attenuation of the seismic waves of the two events is similar, the amplitude difference of the waves recorded at the station indicates the difference in strength of the events. Figure 2.37 shows that the Wenchuan earthquake was stronger than the Tangshan earthquake.

Surface waves of the Wenchuan earthquake took about 6 minutes to travel to Beijing (1500 km away). Some of the regional seismic stations recorded surface waves only, because of high background noise. This caused surface waves (Love waves and Rayleigh waves) with slow propagation velocities to be mistaken for body waves (P and S waves) with fast propagation velocities. Thus a false location was issued 6 minutes after the Wenchuan earthquake: a earthquake of magnitude 4 occurring in the vicinity of Beijing. With more data, this mistake was quickly corrected. This false location determination is a reminder that close attention needs

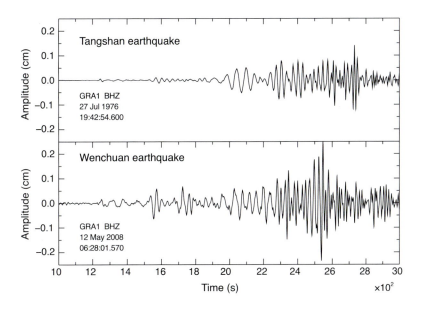

Figure 2.37 Waveforms of the 2008 Wenchuan earthquake and 1976 Tangshan earthquake recorded by the German station GRA1. The maximum amplitude of ground shaking due to Wenchuan is about twice that of Tangshan.

to be paid to the discrimination of seismic waveforms, and quality, not quantity, is the most important consideration in building seismometer stations.

References

Burchfiel B C, Royden L H, Van Der Hilst R D, et al. A geological and geophysical context for the Wenchuan earthquake of 12 May 2008, Sichuan, People's Republic of China. GSA Today, 18(7): 4–11.

Chen J, Hayes G. 2008. Finite fault model—preliminary result of the May 12, 2008 M_W 7.9 eastern Sichuan, China earthquake. http://earthquake.usgs.gov/earthquakes/eqinthenews/2008/us2008ryan/finite_fault.php. Accessed 23 Mar 2011.

Chen Q F, Wang K L. 2010. The 2008 Wenchuan earthquake and earthquake prediction in China. Bull Seismol Soc Amer, 100(5B): 2840–2857.

Chen Y, Li L, Li J, et al. 2008. Wenchuan earthquake: Way of thinking is changed. Episodes, 31(4): 374–377.

Clark M K, Royden L H. 2000. Topographic ooze: Building the eastern margin of Tibet by lower crustal flow. Geology, 28(8):703–706.

Division of monitoring and prediction, CEA. 2009. Scientific Research Report of Wenchuan 8.0 Earthquake. Beijing: Seismological Press. (in Chinese)

Elliott J R, Walters R J, England P C, et al. 2010. Extension on the Tibetan Plateau: recent normal faulting measured by InSAR and body wave seismology. Geophys J Int, 183(2): 503–535. doi: 10.1111/j.1365–246X.2010.04754.x

Huang Y, Wu J P, Zhang T Z, et al. 2008. Relocation of the M8.0 Wenchuan earthquake and its aftershock sequence. Sci China Ser D-Earth Sci, 51(2): 1703–1711.

Hubbard J, Shaw J H. 2009. Uplift of the Longmenshan and Tibetan Plateau, and the 2008 Wenchuan (M = 7.9) earthquake. Nature, 458:194–197.

Jiang H K, Li M X, Wu Q, et al. 2009. Characteristics of the May 12, 2008 Wenchuan M_S8.0 earthquake sequence and discussion on relevant problems. Earthquake Res China, 23(1):34–47.

Kirby E, Whipple K, Harkins N. 2008. Topography reveals seismic hazard. Nat Geosci, 1(8): 485–487.

Lei J, Zhao D P, Su J R, et al. 2009. Fine seismic structure under the Longmenshan fault zone and the mechanism of the large Wenchuan earthquake. Chin J Geophys, 52(2):339–345.

Meng G J, Ren J W, Wang M, et al. 2008. Crustal deformation in western Sichuan region and implications for 12 May 2008 M_S 8.0 earthquake. Geochem Geophys Geosyst, 9(11), Q11007. doi:10.1029/2008GC002144

MIT Report. 2008. Earthquake near Wenchuan, West Sichuan, China 2008 May 12 06:28:01 UTC; Magnitude 7.9. http://quake.mit.edu/~changli/wenchuan.html. Accessed 23 Mar 2011.

Park J, Anderson K, Aster R, et al. 2005. Global seismographic network records the great Sumatra-Andaman earthquake. Eos Trans AGU, 86(6): 57–64.

Shu W, Peng Y J, Zheng Y J, et al. 2002. Crust and upper mantle shear velocity structure beneath the Tibetan Plateau and adjacent areas. Earth Sci, 23(3):193–200. (in Chinese)

Wang W M, Zhao L F, Li J, et al. 2008. Rupture process of the M_S 8.0 Wenchuan earthquake of Sichuan, China. Chin J Geophys, 51(5):1403–1410. (in Chinese)

Wei W, Sun R M, Shi Y L. 2010. P-wave tomographic images beneath southeastern Tibet: Investigating the mechanism of the 2008 Wenchuan earthquake. Sci China Ser D-Earth Sci, 53(9):1252–1259.

Wen X Z, Ma S L, Xu X W, et al. 2009a.Historical pattern and behavior of earthquake ruptures along the eastern boundary of the Sichuan–Yunnan faulted-block, southwestern China. Phys Earth Planet In, 168(1/2):16–36.

Wen X Z, Zhang P Z, Du F, et al. 2009b. The background of historical and modern seismic activities of the occurrence of M8.0 Wenchuan, Sichuan, earthquake. Chin J Geophys, 52(2): 444–454. (in Chinese)

Wikipedia. 2008. 2008 Sichuan earthquake. http://en.wikipedia.org/wiki/2008_Sichuan_earthquake. Accessed 23 Mar 2011.

Zhang Y, Feng W P, Xu L S, et al. 2008. Spatio-temporal rupture process of the 2008 great Wenchuan earthquake. Sci China Ser D-Earth Sci, 53(2): 145–154.

Zheng Y, Ma H S, Lu J, et al. 2009. Source mechanism of strong aftershocks ($M > 5.6$) of the 2008/05/12 Wenchuan earthquake and the implication for seismotectonics. Sci China Ser D-Earth Sci, 52(6):739–753.

Prediction efforts prior to the Wenchuan earthquake 3

- 124 / Earthquake monitoring system in China
- 130 / Prediction efforts prior to the Wenchuan earthquake
- 140 / Development of earthquake prediction research in China
- 152 / Future prospects for earthquake prediction

Many large earthquakes occured in the history of China (Figure 3.1). Chinese seismologists have engaged in earthquake prediction research and practice for half a century, but they failed to predict the Wenchuan earthquake.

Why is it that the Wenchuan earthquake could not be predicted? We believe that the failure was principally due to scientific factors. The Haicheng earthquake prediction in 1975 was basically performed on an empirical basis and even today, after more than three decades of intense research worldwide, our physical understanding of the earthquake rupture process is inadequate to predict any major earthquake with certainty.

To address this question we will first discuss

Figure 3.1 The Pingluo (Gansu Province) earthquake of 1739, caused great ground rupture; the Great Wall was offset 4 m by the earthquake fault (courtesy of Wang Lanmin).

the earthquake prediction efforts at Wenchuan, then we provide a historic perspective on the development of earthquake research in China, finally, we will summarize the principal scientific challenges to earthquake prediction.

3.1 Earthquake monitoring system in China

National earthquake monitoring network

In May 1983, China Earthquake Administration (CEA) and the United States Geological Survey (USGS) began the overall design of the China Digital Seismograph Network (CDSN). By 1986, the first national-level digital seismograph network in China had come into existence, then comprising 9 digital seismic stations (Beijing, Sheshan, Mudanjiang, Hailaer, Urumqi, Qiongzhong, Enshi, Lanzhou and Kunming), and a CDSN maintenance center and data management center. Lhasa and Xi'an digital seismic stations were set up respectively in 1991 and 1995, and joined the seismograph network. CEA and USGS jointly carried out remodeling of CDSN between 1993 and 2001, making its hardware and software systems conform with the technical regulations of the Global Seismographic Network (GSN) established by the Incorporated Research Institutions for Seismology (IRIS). CDSN is now one of the major partners of the GSN (Zhou et al., 1997).

CEA began to establish the China Digital Seismological Observation System in 1996, following a principle of uniform distribution of seismic stations while ensuring intensive observation in some key areas. The observation system was designed to consist of a national digital seismograph network, regional digital seismograph networks and mobile digital seismograph networks. It was completed and began operation at the end of 2000.

The national digital seismograph network is an earthquake monitoring network with uniformly distributed seismic stations all over China. The network consists of 145 seismic stations (equipped with ultra-broadband (UBB) and very-broadband (VBB) seismometers), a data management center and a data backup center. The 145 national seismic stations include 48 primary stations and 97 newly built ones equipped with VBB digital seismographs (Figure 3.2). The average distance between seismic stations is about 250 km (except on the Tibetan Plateau, Figure 3.3). All of the

Chapter 3 · Prediction efforts prior to the Wenchuan earthquake | 125

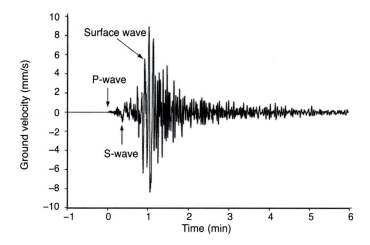

Figure 3.2 Seismogram of Wenchuan earthquake at Kunming station in the national digital seismograph network, located about 600 km south of the epicenter of Wenchuan earthquake. Most station seismograms were off-scale within 500 km of the epicenter.

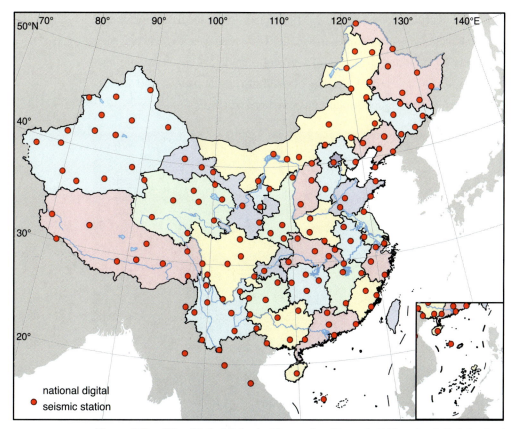

Figure 3.3 China Digital Seismic Observation Network (145 national digital seismic stations).

seismic stations perform 24-bit data acquisition and the waveform data are transmitted in real time to the CDSN center via a satellite network. The 145 stations are located at good sites for observation; most are located in mountain tunnels.

The total number of regional seismic stations on the Chinese mainland has now reached 792, including the 107 seismic stations already built in the capital area close to Beijing. At present, each of the 31 provinces, autonomous regions and municipalities on the Chinese mainland has a regional digital seismograph network, thus there are 31 regional digital seismograph networks in China. The average distance between regional seismic stations is as short as 30–60 km in most provinces, and around 100–200 km in Xinjiang and the Tibetan Plateau. The waveform data are transmitted in real time to the centers of the local seismograph networks.

The mobile digital seismograph networks have 800 portable seismographs. Mobile networks are mainly used to monitor possible foreshocks before large earthquakes, and to investigate and assess local seismicity and source characteristics after large earthquake. They contribute to high-precision hypocenter location and the study of regional seismic activity. They are also used for scientific research. CEA has 600 Güralp seismometers, 600 Reftek-130B data collectors and 600 solar power systems (Figure 3.4).

Figure 3.4   600 mobile digital seismographs are available for monitoring and research.

Other geophysical observations

Observation of changes in various geophysical fields before earthquakes, and exploration of their relationship with the earthquake occurrence, are important aspects of earthquake prediction research in China. Thus in the past 30 years, 1855 precursor stations have been established, to provide observations of different phenomena which include crustal deformation, magnetic field, electrical field, groundwater chemistry, well water level, gravity field, electromagnetic field, stress/strain field etc (Table 3.1). In addition, crustal movement observation network was established since 2000 (Figure 3.5). All of the observed data are sent to China Earthquake Network Center (CENC) from the precursor stations (Figure 3.6).

Table 3.1 Numbers of CEA stations designated to monitoring earthquake precursory anomalies in April, 2008 (Chen et al., 2009)

Crustal deformation	Magnetic field	Electrical field	Aquifer chemistry	Well water level	Gravity field	Stress/ Strain field	Electromagnetic field
358	255	109	493	504	24	76	36

Data processing and analysis procedures

CENC takes responsibility for seismic data collection and transmission, rapid earthquake information dissemination, compilation of the earthquake catalogue, seismic data management and service, and the supervision and technical management of seismograph networks.

CENC collects continuous waveform data in real time from the 145 national digital seismic stations. It can also manage the quasi real-time collection of data from 792 regional digital seismic stations and seismic waveform data from 77 GSN stations via USGS/NEIC. The regional earthquake network centers are capable of quasi real-time collection of waveform data from selected seismic stations of nearby regional seismograph networks via CENC, with a time delay of no longer than 5 seconds.

Using data from both national and regional seismic stations, CENC is capable of detecting $M_L 2.5$ earthquake events in most areas in mainland China. In most of North China, northeastern China, Central China, some

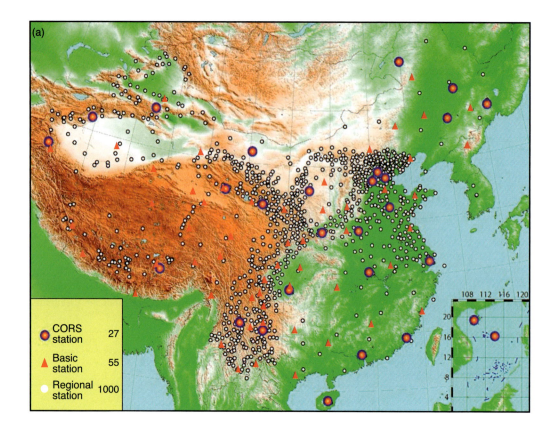

parts of northwestern China and eastern coastal areas, it can even detect events of $M_L 2.0$; in key areas for earthquake surveillance and protection, and in heavily populated major cities, it can detect events with a magnitude as small as $M_L 1.5$. The combined use of GSN and CENC data greatly increased the location accuracy and speed of information dissemination for earthquakes close to the Chinese border or outside it. CENC can release initial information on the epicenter of an $M_S \geqslant 4.5$ earthquake inside or close to China within 10 minutes, and accurate information on the epicenter within 20 minutes. It can release information on an $M_L \geqslant 3$ event located within a regional digital seismograph network in no more than 10 minutes, and information on the focal mechanism solution of an $M_S \geqslant 4.5$ event inside China in no more than 30 minutes.

CENC has set up a seismic data management and service system, which boasts fairly advanced technology, in order to satisfy the requirements of

Chapter 3 · Prediction efforts prior to the Wenchuan earthquake | 129

Figure 3.5 (a) GPS Crustal Movement Observation Network of China, including 27 Continuously Operating Reference Stations (CORS) stations, 55 basic stations and 1000 regional stations. (b) Station details. GPS measurements are made every minute at CORS stations, once a year at basic stations, and at longer intervals at regional stations.

Figure 3.6 China Earthquake Network Center (CENC). All seismic data acquisition is done by CENC.

various users. Users can download seismic waveform data, earthquake catalogues, seismic phase data and focal mechanism data derived from the national digital seismograph network and the regional digital seismograph networks via: http://data.earthquake.cn and http://www.csndmc.ac.cn (Liu et al., 2007).

Data processing is carried out in order to search for earthquake precursors. It is important first to understand the normal background seismic activity, and the normal changes of the various geophysical fields, and then find and examine abnormal changes before past earthquakes (using case studies). Finally, today's observed anomalies can be compared with past anomalies and an opinion on prediction can be proposed.

Over the years, earthquake prediction in China has taken to follow a "four-stage" strategy, that is, long-term (10 years), intermediate-term (1–2 years), short-term (3 months), and imminent (10 days). The prediction of location and magnitude is expected to be progressively refined through the four stages. Although it is too simplistic to characterize earthquake preparation and possible associated precursory anomalies using these timescales, the strategy has been adopted since the 1970s.

The Prediction Regulations for China state that the administrative units of seismological institutions at national and provincial levels "shall organize conferences on the likelihood of future earthquakes to comprehensively analyze and research various earthquake forecasting opinions and abnormal phenomena related to earthquakes, and to formulate earthquake prediction opinions". In practice, the frequency and scope of these conferences are defined by convention. A conference may form a prediction opinion for the following week, month, or year, depending on whether it is a weekly, monthly, or annual conference, respectively.

3.2 Prediction efforts prior to the Wenchuan earthquake

Geophysicists knew that the rugged mountains of Sichuan Province were primed for a "big one". But they didn't know when, or which fault would give way first (Stone, 2008). Two years before, the Institute of Geology of China Earthquake Administration (IGCEA) deployed 300 broadband seismometers—the densest seismic array in the world—around Anninghe, which has been shifting to the east about 10 millimeters a year as the Indian subcontinent pushes the Tibetan Plateau against the Sichuan Basin (Figure 3.7).

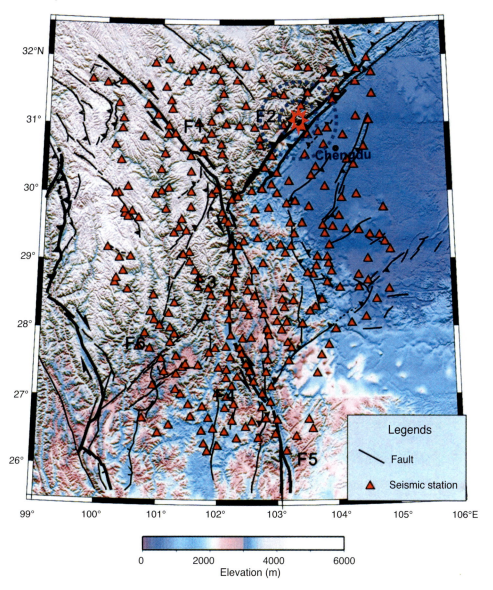

Figure 3.7 Two years before the Wenchuan earthquake, a network of 300 seismometers was deployed around the Anninghe fault (F3) south of Longmenshan fault (F2), where a GPS survey detected great crustal deformation. Only a few seismometers were deployed in the Longmenshan region, because of relatively small deformation there. The red star represents the epicenter of the Wenchuan earthquake (courtesy of Zhang Peizhen).

Seismic network of 300 seismometers

On 12 May 2008, a complex fault system ruptured under the Longmenshan, northeast of Anninghe, releasing an energy equivalent to about 2000 Hiroshima-size atomic bombs. Land to the west of the Longmenshan fault zone had been edging eastward toward the Sichuan Basin at a rate of only a few millimeters per year, according to GPS measurements. Many seismologists believe that the GPS readings blinded researchers to the real threat, and have declared "We did not imagine such a big event happening in Longmenshan."

Before the Wenchuan earthquake, most seismologists perceived two immediate threats. One was the Anninghe fault (F3 in Figure 3.7) — which has a 90 kilometer seismic gap, an eerily quiet stretch with few tremors. The other was the Xianshuihe fault (F1 in Figure 3.7), which runs southwest to northeast, forming a "V" with the Longmenshan fault, and which, like Anninghe, was moving about 10 millimeters per year. The fact that the Longmenshan fault gave way before the others is a challenge to the traditional idea of active fault segmentation (Figures 3.8, 3.9).

Big shock, small GPS displacement

Why did the great Wenchuan earthquake occur at a place where ground displacement, as indicated by GPS survey, is relatively small? Many people think that the greater the displacement, the more active (the more dangerous) the tectonics. The Wenchuan earthquake shows us that this is not true. During the preparation process of an earthquake, a part of the fault system locks, and stress accumulates, but ground deformation is small, and the locked part will probably be the imminent earthquake epicenter. Thus it is better to understand the earthquake process by using spatial analysis of the local displacement. For example, in breaking a chopstick with two hands, deformation of the middle of chopstick is relatively small until the break — in other words, the break does not correspond to the zones of large deformation. If we imagine that a chopstick represents the North-South seismic belt, with Longmenshan located in the middle, it is easy to understand why there may be a big shock but only small ground displacement (Figure 3.10).

Chapter 3 · Prediction efforts prior to the Wenchuan earthquake   133

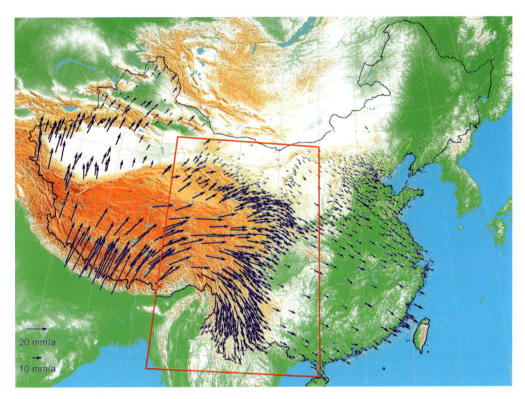

Figure 3.8 In the 1990s, CEA organized a Crustal Movement Observation Network project to obtain the first observations of ground displacement over mainland China. From 1999, GPS observations showed that the displacement at the Longmenshan fault, relative to the stable South China Block, is very small.

Figure 3.9 Enlargement of the Longmenshan area from Figure 3.8.
F1—Xianshuihe fault, GPS rate: 10–12 mm/year.
F2—Longmenshan fault, GPS rate: 1–3 mm/year.
F3—Anninghe fault, GPS rate: 8–10 mm/year.

134 | The Wenchuan Earthquake of 2008

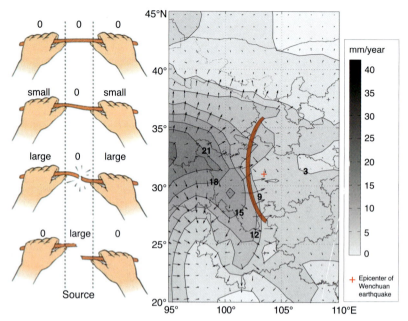

Figure 3.10 If we break a chopstick with two hands, deformation of the middle of the chopstick is relatively small until the break. In other words, the break does not occur where there is largest deformation.

Seismicity pattern

Broadband stations of the digital seismograph network and local network stations in Sichuan and surrounding provinces began routine operation in 1970 (Figure 3.11).

Earthquakes occurring in the area of an ensuing mainshock are called foreshocks, and they are the most convincing precursors for some large earthquakes, such as the 1975 Haicheng earthquake. However, the Wenchuan earthquake, like the Tangshan 1976 earthquake, had no foreshock sequence.

Not only were there no recognizable foreshocks, the long-term seismicity along the Longmenshan fault zone, recorded from 1970 when seismic networks began routine operation in Sichuan and its neighbouring areas, showed no anomalous temporal pattern prior to the Wenchuan earthquake (Figure 3.12).

Many earthquakes were recorded in Sichuan and surrounding areas. Figure 3.13 shows the epicenter distribution of earthquakes of magnitude 4 or above in Sichuan and surrounding areas since 2001. From the spatial distribution, there was no sign of increased seismic activity near Wenchuan.

Seismologists paid special attention to the Longmenshan fault from the 1970s, geologists studied its active tectonics for several decades, and surveyors made GPS observations from the 1990s. All of them underestimated the seismic risk of the Longmenshan fault. Maybe this is because there were no historical earthquakes with magnitude greater than 7. In addition, many scientists thought that the Xianshuihe and Anninghe faults were more dangerous than the Longmenshan fault. It can be seen that

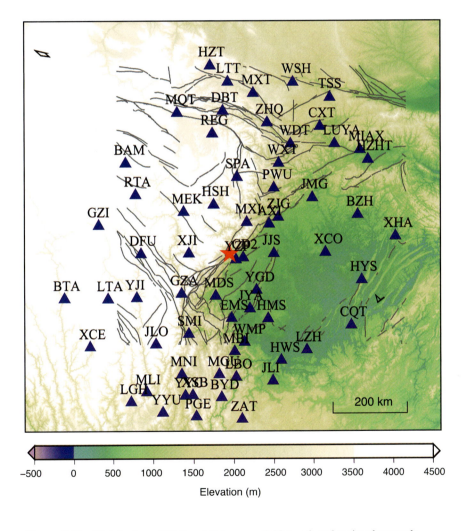

Figure 3.11 Distribution of National Seismograph Network and regional network broadband stations in Sichuan and surrounding provinces. Solid triangles show station locations; the star represents the epicenter of the 12 May 2008 $M_S 8.0$ Wenchuan earthquake. Grey lines represent Quaternary active faults (from Zheng et al., 2009).

136 | *The Wenchuan Earthquake of 2008*

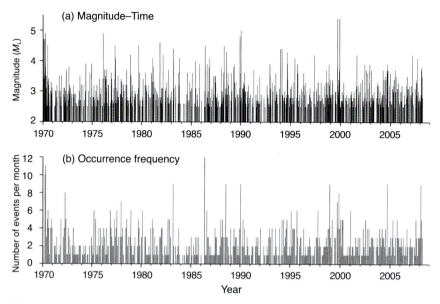

Figure 3.12 Time sequences of earthquake occurrence along the Longmenshan fault zone (roughly in the area of Figure 3.11) before the Wenchuan earthquake. They begin in 1970, when seismic networks began routine operation in Sichuan and its neighbouring regions (catalogue complete for $M_L \geq 2.5$). (a) Magnitude vs. time. (b) Frequency of occurrence vs. time, using a counting bin of 1 month.

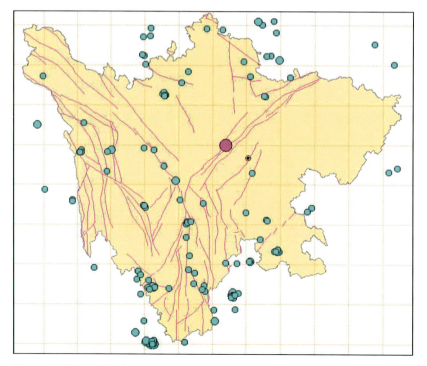

Figure 3.13 Distribution of earthquakes of magnitude 4 or above in Sichuan and surrounding areas since 2001. The red circle represents the Wenchuan earthquake.

in Figure 3.13 there is a seismic gap in the Longmenshan region, which has lasted since 1977. It is possible that this gap resulted from fault-locking and strain energy was accumulating there for the last 30 years at least.

According to Wang's study (Wang et al., 2006) of various reported prediction cases, especially the 1975 Haicheng earthquake, other frequently mentioned types of anomalies have not been useful for short-term and imminent predictions, but their appearance may be related to future earthquake activity. It is therefore worth investigating the presence or absence of these anomalies prior to the Wenchuan earthquake, since the system for anomaly monitoring and reporting in 2008 was much simpler and better organized than ever before.

Precursor anomalies

A simple and objective measure of the temporal pattern of anomalies known to earthquake workers in Sichuan can be obtained from two sets of data: (1) the number of anomalies discussed at weekly provincial prediction conferences, and (2) the number of "macroscopic" anomaly reports submitted to the Sichuan Provincial Seismological Bureau in Chengdu by other individuals or organizations. "Macroscopic" anomalies are those that can be observed without sophisticated instruments. The weekly conference would discuss all types of anomalies, including the "macroscopic" anomalies. Precursor data are shown in Figure 3.14 from 2002 through May 2008. No alarming temporal pattern can be identified prior to the Wenchuan earthquake, and the anomalies occur with or without earthquakes (Figure 3.14). In Table 3.2, the reports used for Figure 3.14 are reorganized into several categories for each year. Earthquake workers at the provincial bureau are required to conduct site visits and interviews to validate each macroscopic anomaly report. The original and validation reports are filed in their headquarters in Chengdu. Unfortunately, we were unable to locate the validation reports for 2002–2005, but this does not affect the conclusion that there is no alarming temporal pattern.

In order to determine whether there could have been any indicative spatial pattern of reported anomalies over a broad region, Chen et al. (2010) show in Figure 3.14 the sites of anomalies discussed at weekly prediction conferences of the seismological bureaux of Sichuan and its neighbouring provinces within a month prior to the Wenchuan earthquake. The distribution does not show any pattern that would draw attention to the area of the Wenchuan earthquake (Figure 3.15).

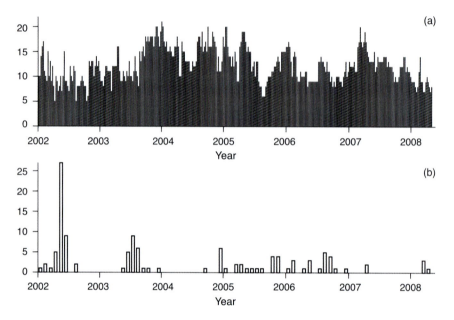

Figure 3.14 Time sequences of "anomalies" before the Wenchuan earthquake from 2002. (a) Number of anomalies discussed at each weekly prediction conference of the Sichuan Provincial Seismological Bureau. (b) Number of macroscopic anomalies reported to the provincial bureau each month (from Chen et al., 2010).

Table 3.2 Numbers* of macroscopic anomaly reports received by the Sichuan Provincial Seismological Bureau prior to the 12 May 2008 Wenchuan earthquake (Chen et al., 2010)

Year	Ground-water	Unusual sighting**	Electro-magnetic	Animal behaviour	Plant peculiarity	Weather pattern	Annual total
2002	37/?	3/?	0	6/?	0	1/?	47/?
2003	16/?	3/?	0	4/?	0	1/?	24/?
2004	6/?	0	0	0	1/?	0	7/?
2005	9/?	1/?	0	7/?	0	0	17/?
2006	15/13	2/1	0	1/0	2/2	0	20/16
2007	2/1	0	0	0	0	0	2/1
2008	3/3	0	0	1/0	0	0	4/3

* For each type of anomaly, the first number is the number of reports received, and the second number is the number of reports validated to be real. Validation reports for 2002–2005 are missing.
** Unusual sightings include sighting a strange light, smoke, fire, etc. or hearing an unusual sound.

Chapter 3 · Prediction efforts prior to the Wenchuan earthquake | 139

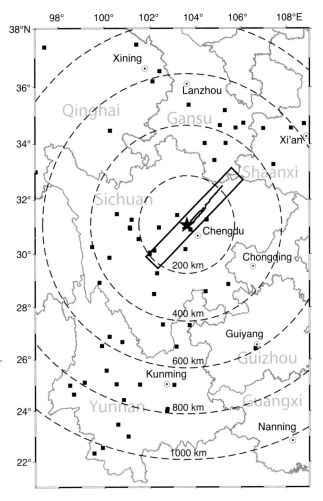

Figure 3.15 Locations of all types of anomalies discussed at provincial weekly prediction conferences held by the seismological bureaux of Sichuan Province and neighbouring provinces or autonomous regions during a one-month period immediately preceding the Wenchuan earthquake. Dashed lines show distances from the epicentre (star). The surface rupture of the Wenchuan earthquake (solid lines) is derived from Xu et al. (2008). The rectangular box indicates the Longmenshan fault zone. Names of the provinces or autonomous regions are in grey, and their capital cities are in black (from Chen et al., 2010).

The number of reported anomalies prior to the Wenchuan earthquake is much smaller than what was reported for the Haicheng earthquake (Wang et al., 2006). It is not clear whether the discrepancy reflects a difference between the physical processes of these earthquakes or different degrees of attention paid to precursor monitoring. The Haicheng earthquake occurred during a period of very strong earthquake activity over large regions of China, but the Wenchuan event had a quieter background seismicity. Perhaps their preparation processes are sufficiently different to give rise to different amounts of precursory anomalies (Chen et al., 2010).

The primitive state of earthquake prediction, says a senior CEA geophysicist, is similar to that of weather forecasting a century ago, when people relied on observations of the sky and animal behavior. "Meteorologists have made the transition from empirical to physical prediction," he says, "We haven't."

3.3 Development of earthquake prediction research in China

Severe earthquake disasters in China

Throughout history, natural disasters, particularly earthquakes, have exacted a heavy toll in death and human suffering.

China is situated in the southeastern part of the Eurasian plate and is bordered by the western Pacific seismic belt in the east and the Himalayan-Mediterranean seismic belt to the west. It has been shown in section 2.3 that it is also a zone of strong intraplate seismicity (Figures 3.16, 3.17). Thus China is particularly vulnerable to great and frequent earthquakes. Since the beginning of the 20th century it has suffered 660 earthquakes over magnitude 6, and of these 99 were of magnitude 7 to 7.9, and 11 of magnitude 8 or above. The distribution of earthquakes in China spans a wide area; 21 out of 31 provinces have suffered earthquakes above magnitude 6. Western and the northeastern China have been particularly prone to great earthquakes.

Great earthquakes can severely disrupt community lifelines—the systems that provide food distribution, water supply, waste disposal, and communication locally and with the rest of the China. For example, the Tangshan earthquake of 1976 destroyed a major industrial city in a moment (Figure 3.18) and killed over 240,000 people and injured many more (Chen et al., 2002).

China is not the country with the largest earthquake in the world (Figure 3.19), but it is the most earthquake disaster-prone country in the world. This is due to (1) economic backwardness, (2) poor quality construction, (3) poor disaster awareness, and (4) mistaken popular belief that precautionary action is wholly the responsibility of the government (previously the emperor). To better understand earthquake prediction in China, it is necessary to know the cultural/economic and social traditions.

地裂泉涌中有魚物或城郭房屋陷入地中或平地突
成山阜或一日數震或累日震不止河渭大泛華岳終
南山鳴河清數日官吏軍民壓死八十三萬有奇三十
七年正月庚申陝西地震三月丁醜昌平州地震五月
丁卯蒲州地震三日聲如雷六月甲申又震十月丙
午華州地震聲如雷至壬子又震戊午復大震傾陷廬
捨甚多三十八年七月辛巳南京地震有聲三十九年
四月嘉興湖州地震屋廬搖動如帆河水撞激魚皆躍
起四十年二月戊戌甘肅山丹衛地震有聲壞城堡廬
捨六月壬申太原大同榆林地震寧夏固原尤甚城垣
墩臺府屋皆摧地涌黑黃沙水壓死軍民無算壞廣武
紅寺等城四十一年正月丙申京師地震是歲寧夏地
震圮邊墻四十五年正月癸巳福建福興泉三府同日

〈明史卷三十　志　二十〉

Figure 3.16 This historical record indicates that in 1556 (Ming dynasty of China), a great earthquake occurred in the middle of China (Shaanxi Province), which killed 830,000 people. This earthquake in Hausien, Shaanxi Province, China in the morning of 23 January 1556 caused the worst natural disaster in recorded history—at least in terms of lives lost. In Chinese historical records, this event is often referred to as the "Jiajing Great Earthquake" because it occurred during the reign of Emperor Jiajing of the Ming dynasty. The "Shaanxi Earthquake" as it became later known, had an estimated magnitude ranging from 8.0 to 8.3 on the Richter scale (assigned moment magnitude is 8), and had an estimated intensity of XI on the Modified Mercalli Intensity scale. Its epicenter was near Mount Hua in Shaanxi, close to present day Weinan City. The earthquake was responsible for the devastation of 98 counties and eight provinces in Central China, and was particularly destructive in Shaanxi Province. The destruction extended over an area of 800 km. In some of the counties, the average death toll was estimated to be about 60 percent of the population. According to historical records, a total of 830,000 people lost their lives, most from the collapse of poorly constructed houses and of loess cave dwellings.

Figure 3.17 At 20:00 on 16 Dec 1920, a M8.5 earthquake occurred in Haiyuan of western China, causing 200,000 casualties. The willow tree at Shaomaying was split by the earthquake (courtesy of Haiyuan Earthquake Museum).

Figure 3.18 Tangshan earthquake (M7.8) of 1976 destroyed a major industrial city Tangshan in a moment and killed 240,000 people. Heavy damage sustained by a part of Tangshan City located within the intensity X zone.

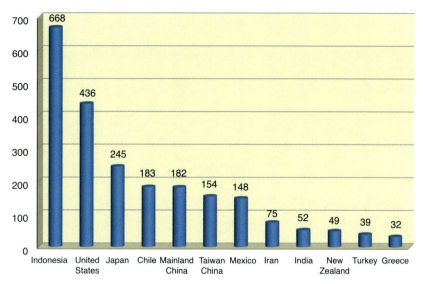

Figure 3.19 The number of earthquakes with magnitude 6 or above in different countries. China does not have most of these earthquakes, but it is the most earthquake disaster-prone country in the world. Data source: USGS earthquake catalogue (1 Jan 1964 to 31 Dec 1998).

Organized efforts for earthquake prediction (1966)

Organized efforts for earthquake prediction in China started in March 1966. On 8 and 22 March 1966, earthquakes of magnitudes 6.8 and 7.2 struck Xingtai, Hebei Province and caused immense devastation to the area. Both earthquakes caused damage in more than 100 neighboring counties, and 8000 deaths.

The Xingtai earthquake was the first major earthquake that had occurred in densely populated eastern China since the founding of the People's Republic of China in 1949. The extensive damage it caused and the close proximity of its epicenter to Beijing, alarmed the central government. On the same day, Premier Zhou Enlai went to the disaster area and announced the start of an earthquake prediction program in China (Figure 3.20).

3600 researchers from over 100 government research institutions and universities immediately gathered in the epicentral area and began a massive prediction campaign. Some of the researchers felt that they had found clues to precursory anomalies which they could use to make short-term predictions. Well known anomalies include temporal and spatial patterns of small earthquakes, groundwater changes, and aberrant animal behaviour.

Figure 3.20 The Xingtai earthquakes were to be a turning point for earthquake prediction research in China. Premier Zhou Enlai (middle of front row) toured the area three times and showed great concern. He challenged the current generation of scientists to solve the earthquake prediction problem. Under his leadership China embarked on a national campaign for systematic investigation and research on earthquake prediction.

The acclaimed three-stage formula of "many foreshocks, short hiatus, large earthquake" was proposed at this time.

China Earthquake Administration (CEA)

Eight damaging earthquakes of $M > 7$ occurred in 1966–1976 in central and eastern China (Figure 3.21), emphasizing that there was an urgent need for comprehensive earthquake hazards reduction, and driving forward the prediction program.

In 1971, the State Seismological Bureau (SSB) was formed (later becoming China Earthquake Administration) with the primary task of predicting earthquakes. In 1975, the SSB became an entity under the State Council reporting directly to the Premier, similar in status to a ministry, thus emphasizing that earthquake prediction was a government mandate, and not merely a scientific program (Figures 3.22, 3.23).

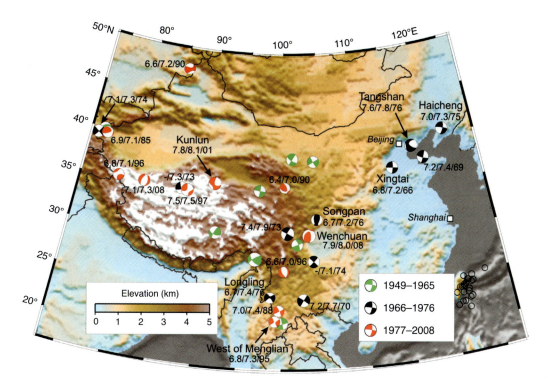

Figure 3.21 Map showing $M_s > 7$ earthquakes in mainland China from October 1949 to May 2008. Focal mechanisms shown for events before 1976 are first-motion solutions based on Zhang et al. (1990) and for later events are centroid moment tensor solutions based on the GCMT Project (http://www.globalCMT.org). Events after 1966 are labelled M_W/M_S/Year and some are named (from Chen et al., 2010).

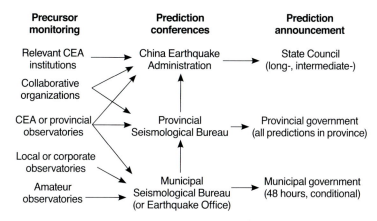

Figure 3.22 Schematic illustration of the reporting and announcement system of Chinese earthquake prediction. Collaborative organizations include universities and other ministries. "Provincial" pertains to provinces, province-level cities, and autonomous regions. "Municipal" pertains to cities, prefectures, and counties. In addition to the regular reporting system, any individual can submit prediction opinions directly to CEA (from Chen et al., 2010).

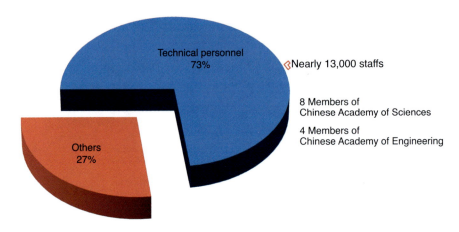

Figure 3.23   Allocation of CEA's 13,000 staff.

Haicheng earthquake prediction

On 4 February 1975, an earthquake of magnitude 7.3 struck Haicheng County, Liaoning Province—a densely populated area of eastern China. In this situation, over 10,000 deaths might be expected, but this earthquake actually caused 2041 casualties, because an earthquake prediction was issued before the mainshock. Measures taken by various levels of government in this area on 4 February 1975 indeed saved thousands of lives.

Wang et al.(2006) studied declassified Chinese documents and interviewed key witnesses. They reported that there were two official middle-term predictions but no official short-term prediction. On the day of the earthquake, a county government issued a specific evacuation order, and actions taken by provincial scientists and government officials also effectively constituted an imminent prediction. These efforts saved thousands of lives. Evacuation was extremely uneven across the disaster region, and critical decisions were often made at very local levels.

The most important precursor was a foreshock sequence. Foreshocks are precursors that are clearly physically related to the occurrence of mainshocks, and so the usefulness of foreshocks for prediction, particularly for imminent earthquake prediction, is particularly important.

From 1 February 1975, 521 foreshocks were recorded at Shipengyu seismic station located about 20 km from the Haicheng earthquake epicenters (M_S = 7.3, 4 February 1975). Several days before the mainshock and near its eventual epicenter the earthquake activity began to increase; this is the common feature of foreshocks (Chen, 1979). The Haicheng foreshock sequence was consistent with the three-stage formula "many foreshocks, short hiatus, large earthquake" proposed after the Xingtai earthquake of 1966.

We use the time differences between t_s (arrival time of S-wave) and t_p (arrival time of P-wave) obtained from the nearest station to study the spatial clustering of foreshocks. Figure 3.24 shows the distances from earthquake

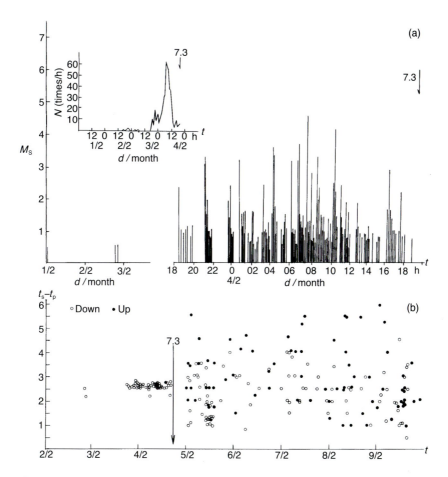

Figure 3.24 (a) Magnitude of foreshocks which occurred near Haicheng during the period 1 to 4 February, 1975. The main shock of magnitude 7.3 struck at 7:36 pm, 4 February. The inset diagram illustrates the average number of foreshocks per hour from 1 until 4 February (Chen et al., 1988). (b) t_s–t_p and senses of first motions of P waves recorded at Shipengyu station during 1 to 9 February 1975; arrow shows the occurrence time of the main shock of Haicheng, M7.3 (from Chen et al., 1999).

epicenters to the station according to the values of t_s-t_p. It is seen in Figure 3.24 that the values were almost the same (t_s-t_p = 2.5 sec). Considering that all the foreshocks were located in the same direction from the station (Figure 3.24), we concluded that the foreshock epicenters were spatially densely concentrated. On the other hand, the epicenters of aftershocks were much more scattered (t_s-t_p ranging from 1 sec to 5.5 sec, see Figure 3.24).

It is very difficult to collect enough data to determine the focal mechanism of each event, because most foreshocks are not large enough. We are interested in the consistency of focal mechanisms of the many foreshocks in foreshock sequences, rather than the exact focal mechanism of each foreshock.

For a given station and epicentral area, we can monitor changes of focal mechanisms by using the directions of P-wave first motions recorded in the station. It can be seen from Figure 3.24(b) that the directions of P-wave first motion at Shipengyu station (the nearest station to the foreshocks) were all downward before the mainshock, which indicated no great change in focal mechanism of the foreshocks. From the 521 foreshocks, 79 first P-wave motions were clearly identified at Shipengyu station, 78 of which had downwards first motions. On the other hand, significant changes in first motion directions occurred for the aftershock pattern, where there was a much smaller proportion of downwards first motions. The first motion directions started to change immediately after the main shock. In the case of the 1995 Menlian earthquake in Yunnan Province (Figure 3.25, Table 3.3), consistency in first motion direction was used to predict whether the largest earthquake had occurred or not. The consistency of focal mechanisms was an important characteristic of the foreshock sequence before the Haicheng earthquake of 1975.

Haicheng did not solve the prediction problem for other earthquakes. However, as the world's first, and so far the only, well-documented case in which an evacuation order was issued for the right place at the right time, just hours before a major earthquake, it deserves a prominent place in human history (Figure 3.26). It is the one event that keeps the hope for precursor-based earthquake prediction alive.

One year later, the Tangshan *M*7.8 earthquake occurred (1976). No short-term prediction was made, and the earthquake caused 240,000 deaths, and severe losses and destruction. Extrapolating from a limited number of

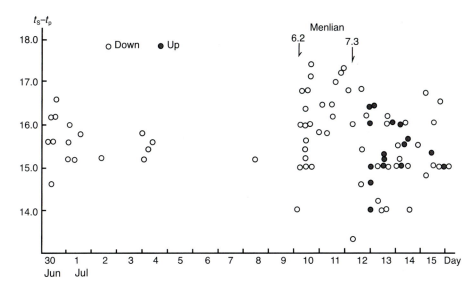

Figure 3.25 Foreshocks of the Menlian earthquake, Yunnan Province, magnitude 7.3. t_s-t_p and first motion directions of P waves were recorded at a seismic station 120 km from the Menlian earthquake epicenter during 1–15 July. Open circles represent downwards first motion of P waves, and solid circles represent upwards first motion. The arrows represent the occurrence times of the largest foreshock and the main shock.

Table 3.3 Foreshock sequences of mainland China during 1966–1996

Earthquake	M	Latitude	Longitude	Foreshock sequence start from	Main shock	Total no.	$M_{mainshock} - M_{foreshock}$	Distance from main shock (km)
Xingtai	6.8	37.18	114.54	1966-03-06	03-08-05h	23	1.5	10
Wushi	6.1	41.19	79.22	1971-03-21	03-24-04h	29	0.1	10
Haicheng	7.3	40.63	122.81	1975-02-01	02-04-19h	521	2.6	15
Menla	5.7	21.18	101.30	1975-10-23	10-28-07h	33	1.0	20
Lunglin	7.3	24.22	98.38	1976-05-29	05-29-23h	16	2.0	20
Ninglang	6.7	27.30	101.05	1976-11-04	11-07-02h	16	3.2	15
Datong	5.8	39.57	113.50	1991-03-21	03-25-02h	13	4.6	10
Menlian	7.3	22.10	99.60	1995-07-10	07-12-08h	125	1.8	15

$M_{mainshock} - M_{foreshock}$ means the magnitude differences between the main shocks and the maximum foreshocks.

government reports of that time, Guo (2008) estimated that in the fall of 1976, about 400 million of a total Chinese population of 930 million spent some nights in temporary earthquake shelters.

No other earthquake in the last century has been as catastrophic or has claimed as many lives as the earthquake that struck the city of Tangshan in

Figure 3.26 The notice at the movie theater reads "tonight's movies show outdoors". The theater manager was convinced by workers at the earthquake observatory that there was an impending earthquake that night and decided to show movies outdoors overnight to attract people away from their houses. The earthquake occurred during the movie show.

northern China on 28 July 1976. Tangshan, a thriving industrial city with one million inhabitants, is located in Hebei Province, about 95 miles east and slightly south of Beijing and about 280 miles southwest of Haicheng. Although the region had experienced moderate seismic activity in the past, this time there were no foreshocks, and no warning.

Over the 32 years from 1976 to 2008, many more earthquakes occurred without being predicted, and countless prediction opinions were submitted to SSB/CEA without being followed by the predicted earthquakes. It is increasingly clear that, whatever its future potential, it is impractical at present for society to rely on predictions to prevent earthquake disaster.

Earthquake Act

While it is heartening to hold on to the good experience of the Haicheng prediction, in the foreseeable future society cannot rely on predictions in order to prevent earthquake disaster. The President of China issued order No. 94 in 1997 (a new version now amended and adopted as Order No. 7 of 2008): The Law of the People's Republic of China on Protecting Against and Mitigating Earthquake Disasters (hereafter referred to as the Earthquake Act, Figure 3.27).

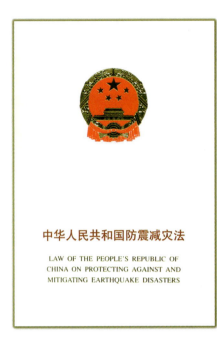

Figure 3.27 *The Law of the People's Republic of China on Protecting Against and Mitigating Earthquake Disasters*, adopted at the 29th meeting of the standing committee of the eighth national People's congress of the People's Republic of China on 29 December 1997, is hereby promulgated and shall come into force as of 1 March 1998. The new amended version was issued in 2008 (see Appendix 3).

From the above brief review, it can be seen that earthquake prediction in China is government-sanctioned and legally regulated.

Enactment of the Earthquake Act in 1998 was an important development. The Earthquake Act regulates (1) monitoring and prediction, (2) preparedness and mitigation, (3) emergency response and rescue, and (4) post-disaster relief and rebuilding. It specifies the essential role of the "competent administrative department of the State Council responsible for seismic work", namely the CEA, in the first three of these activities. In fact, since its establishment, the CEA's work had long gone beyond making predictions. However the Earthquake Act officially codified its multi-functional role, especially its role in preparedness and mitigation. Until then, the seismic design code was more of a policy enforced through administrative orders by various committees and offices; non-compliance might bring disciplinary repercussions but was not punishable by law. The Earthquake Act gave the code legal status.

Lessons learned from the Wenchuan earthquake triggered a substantial revision of the Earthquake Act. One addition to the Act is the requirement that the building codes for densely populated buildings such as schools and hospitals meet higher specifications for seismic protection than those for regular buildings in the same area. There is great hope that seismic risk mitigation will continue to improve across China.

3.4 Future prospects for earthquake prediction

Realistic public expectation of earthquake prediction

What is the capability of earthquake prediction today? How many earthquakes can be predicted? And how many earthquakes cannot be predicted? After nearly half a century of practical experience, these questions need a scientific and objective answer.

We have very limited understanding of the environment, internal processes and the stress-strain state of the Earth's interior, therefore earthquake prediction today has to be an empirical prediction based on past experience.

The most important precursor for prediction is a foreshock sequence. Prediction of the Haicheng earthquake of 1975 was mainly based on foreshocks; this was introduced in section 3.3. The following example describes an earthquake with precursory foreshocks, for which no prediction was issued.

In the early morning of 14 April 2010, a strong earthquake ($M_S7.1$) occurred in Yushu district, Qinghai Province, China (Figure 3.28). More than 2000 lives were lost and over 100,000 injured.

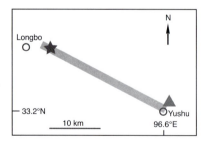

Figure 3.28 Locations of the 14 April 2010 Yushu earthquake (star) and the Yushu Seismograph Station (triangle). The solid line represents the earthquake rupture (from Ni et al., 2010).

Among all short-term precursors to earthquakes, foreshocks are accepted as the best indicator for impending strong earthquakes. A M4.7 event near Yushu a few hours before the mainshock was strong enough to be recorded at a dozen stations and the P-waves had a high signal-noise ratio, making it possible to determine the relative location between the main shock and the M4.7 event. The relative location is found to be within 2 km, and thus the M4.7 event is a genuine foreshock.

At Yushu, the seismicity began about 28 hours earlier than the main shock, and became much higher after the M4.7 event about two hours before the mainshock. The earthquake also seems to become stronger as indicated in Figure 3.29(b). In the halfhour before the mainshock there is almost no seismicity, following the pattern "many foreshocks, short hiatus, large earthquake" proposed after the 1966 Xingtai earthquake. Figure 3.29(c) shows the relative location of e1 (the foreshock at the 12th hour), the M4.7 foreshock, e2 (an earthquake after the M4.7 event), ef (the last foreshock), and the main shock, where the relative locations are inferred from S-P times for waveforms recorded at the Yushu seismograph station. The main shock ruptured about 30 km unilaterally south-eastward towards Yushu City (Ni et al., 2010).

Foreshocks are easy to identify once the largest earthquake, the main shock, has occurred. But before the main shock occurs, it is usually difficult to tell whether an event is a foreshock or the main shock. We compared waveforms recorded at Yushu station for all other events before the main shock, and found that they have very similar waveforms. The similarity of waveforms qualifies these events as a foreshock (Figure 3.30).

The pattern of the Yushu foreshocks seems comparable to Haicheng, though on a much shorter timescale, with increasingly frequent shocks up to a half hour of quiescence that ended with a final foreshock two minutes before the quake (Figure 3.28). Figure 3.29 shows the increase in the magnitude of foreshocks, suggesting that the eventual event might be stronger than the M4.7 event.

We found that there were 159 major earthquakes with magnitude $M > 5.5$ in mainland China during 1966–1996, based on a preliminary determination of the Chinese earthquake catalogue (provided by the Center of Analysis and Prediction, CEA). Among them eight earthquakes had foreshock sequences (Table 3.3). Thus the ratio of main shocks preceded by foreshocks to the

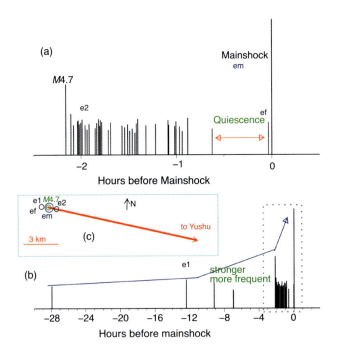

Figure 3.29 Foreshock sequence of the Yushu earthquake. (a) Three hours before the main shock. (b) Thirty hours before the main shock. (c) Relative locations of e1 (the foreshock at the 12th hour), the M4.7 foreshock, e2 (an earthquake after the M4.7 event), ef (the last foreshock), and the main shock (from Ni et al., 2010).

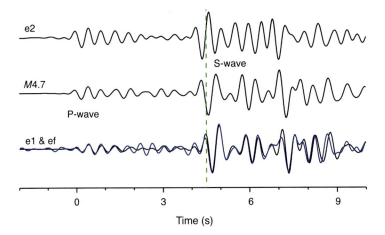

Figure 3.30 Seismographs of the 14 April 2010 Yushu earthquake (recorded at Yushu seismograph station, 25 km from the epicenter). The foreshocks have similar waveforms (aligned on P waves), but with minor difference in S-P interval. The green line is plotted to show the misalignment of S waves. Foreshocks e1 (black trace) and ef (blue trace) have very similar waveforms, suggesting that their locations are very close. The small variation (up to 0.4 s) of S-P interval for the M4.7 and e2 events indicates a foreshock zone size of 3 km (from Ni et al., 2010).

total number of earthquakes of magnitude $M > 5.5$ is about 5%.

In discussing foreshocks, we emphasize two points. First, very few earthquakes have identifiable foreshocks (about 5% in mainland China). The second point is related to the first. Earthquake clustering in space and time takes different forms. In particular, there are occasional foreshock sequences before mainshocks, and much more frequent earthquake swarms, in which no earthquake is sufficiently larger than the others to be classified as a mainshock. Discrimination between foreshock sequences and earthquake swarms, i.e., the recognition of differences of seismicity pattern between foreshock sequences and other sequences not followed by a mainshock, is a difficult problem.

The best capability of prediction of future earthquakes by using foreshocks will be about 5%. Is this estimate is too low for prediction to be useful? We can perform a comparison between earthquake prediction and weather forecasting. The Director-general of China Meteorological Administration, Zheng Guoguang, pointed out in 2007 that the current storm (50 mm/24 hours) prediction success rate is about 19%, while the figure in the United States is similar 22% (http://www.ce.cn/). For different sudden natural disasters, the ability to predict is also different. For some natural disasters it is clear that there will be observational data from which a forecast is capable of being made with a relatively high success rate, such as tsunamis and volcanic eruptions. Forecasts of severe weather will not be as good as these, even though much observational data exists. For earthquakes occurring deep underground, the forecasting capability is worst. Even the prediction success rate for tsunamis, according to the International Pacific Tsunami Centre, is only 25%, giving a false alarm rate of 75% (http://www.tsunami.noaa.gov/tsunami_story.html).

Forecasting ability is very limited for many sudden natural disasters. An earthquake prediction success rate of 5% based on foreshocks, if achievable, is the most objective estimate.

The public expectation in earthquake prediction capability is unreasonably high. What has led to the unrealistic public expectation for earthquake prediction? Has the emphasis on prediction had a negative effect on mitigation? The Wenchuan earthquake has raised these sharp questions; we now need to think deeply and learn from its lessons.

Why is the prediction effort so persistent?

Quite apart from its present scientific status and possible future research, governments of most industrial countries consider earthquake prediction to be presently impractical. Instead, it is commonly accepted that the most effective way of minimizing the impact of earthquakes on human life and the economy is to strengthen our built environment, based on scientific assessment of earthquake hazard and risk.

On the other hand, the philosophy of earthquake prediction in China is fundamentally influenced by Chinese culture and history. The practice of Chinese medicine over millennia demonstrates that it is possible to use an empirically established connection between events to make predictions, without understanding the connection.

To answer "why does China persist with prediction efforts", there are two factors which support persistence in earthquake prediction.

Although China's economy has made great progress, there has been very uneven development between regions. There is a large gap between the eastern and western regions; for example, the per capita GDP of Guizhou Province is only one-tenth that of Shanghai (Figure 3.31).

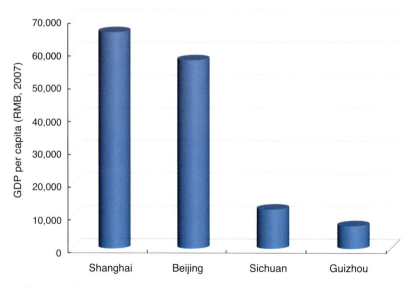

Figure 3.31 China's economic development shows significant regional differences, between eastern and western economic development, as shown by differences in per capita GDP.

Many experts have pointed out that the earthquake hit an area that had been largely neglected and untouched by China's economic rise. Health care is poor in inland areas like Sichuan Province, highlighting the widening gap between prosperous urban dwellers and struggling rural people.

The lack of seismic design for most rural areas of China is blamed on the very poor economic conditions of the time. The contrast between the performance of modern buildings and that of the poor housing in rural districts is heartbreaking (Figure 3.32). It is difficult to change these conditions in the short term. Earthquake prediction is still a means of mitigation; we suggest we should not neglect the 5% chance.

There are many ways to mitigate natural disasters, such as hazard assessment, prediction, and early warning, rescue etc.; relying on a single channel is not enough. For example, various rescue teams (domestic, international, professional) brought out from the

Figure 3.32 Rapid economic growth, but big differences in regions. Modern buildings in large cities: (a) National Stadium (Beijng); (b) Pavilion of China, World Expo (Shanghai); (c) and (d) Old dilapidated houses in Sichuan mountains.

ruins a total of 371 survivors in Wenchuan earthquake.

We do not know how much money was spent on the rescue teams, but certainly it was significant when compared to the total earthquake losses (845 billion yuan). The number of survivors rescued was small, compared to the total number of victims in the disaster. Considering that earthquake prediction (such as in Haicheng) may save tens of thousands of lives, we should not give up those opportunities even if the chance of success is only 5%, considering that research funding would be a small fraction of the cost of the relief effort. Since disaster mitigation requires an integrated program, why should it not include earthquake prediction?

From prediction to mitigation

China's precursor-based earthquake prediction program was initiated in an unusual social environment and has always been a task assigned by the government. For historical reasons, earthquake prediction in China is government-sanctioned and law-regulated. In spite of the installation of massive precursor monitoring networks and an elaborate schedule of prediction conferences, no anomalous pattern was identified before Wenchuan earthquake that would have enabled its prediction.

Compared to the situation 30 years ago, earthquake prediction today suffers more than most from administrative constraints, non-scientific management, and lack of scientific innovation and lively vigor. Earth science has made great progress over the past 30 years, but the level of prediction research has not progressed much. Earthquake prediction is a science, but its management has been performed by administrators.

Frankly speaking, earthquake prediction in China today is prediction based on earthquake catalogs. With no changes to the current level of earthquake prediction and science management, earthquake prediction has no hope.

Friedrich Engels, a German philosopher, said: A clever nation learns from disaster and error much more than usual. Three aspects of prediction research need to be developed with priority:
- medium-term forecast: determine the location and intensity of the future earthquake;
- short-term prediction: determine the occurrence time of the future earthquake;
- move from empirical prediction to physical prediction.

On the other hand, the Wenchuan earthquake has overwhelmingly demonstrated the vital importance of seismic risk mitigation. In contrast to Tangshan 1976, where seismic design was not then required, such design had been required in the Wenchuan area since 1990. Buildings that met the design standard suffered much less damage than those that did not. In practice, progress lagged far behind the promulgation of regulations; a stricter enforcement of anti-seismic design provisions and a wiser selection of construction sites at which the anti-seismic designs were used would have prevented many deaths and greatly reduced the destruction of property. The performance of the prediction program in China has provided useful experience and taught many lessons. The most important lesson is that, regardless of its future potential, it is presently impractical to rely on prediction to prevent earthquake disasters. The practical approach is to strengthen the resilience of our built environment based on an assessment of seismic hazard.

References

Chen Q F, Wang K L. 2010. The 2008 Wenchuan earthquake and earthquake prediction in China. Bull Seismol Soc Amer, 100(5B): 2840–2857.

Chen Y. 1978. Consistency of focal mechanism as a new parameter in describing seismic activity. Acta Geophys Sin, 21(2): 142–159.(in Chinese)

Chen Y, Liu J, Ge H K. 1999. Pattern characteristics of foreshock sequences. Pure Appl Geophys, 155(2–4): 395–408.

Chen Y, Tsoi K L, Chen F B, et al. 1988. The Great Tangshan Earthquake of 1976: An Anatomy of Disaster. Oxford: Pergamon Press.

Guo A. 2008. The Great Earthquake of Tangshan, China. Xi'an: Shaanxi Science and Technology Publishing House. (in Chinese)

Liu R F, Cai J A, Peng K Y, et al. 2007. The project of seismological data sharing. Earthquake, 27 (2): 9–16. (in Chinese)

Ni S D, Wang W T, Li L. 2010. The April 14th, 2010 Yushu earthquake, a devastating earthquake with foreshocks. Sci China Ser D-Earth Sci, 53(6): 791–793.

Wang K L, Chen Q F, Sun S H, et al. 2006. Predicting the 1975 Haicheng earthquake. Bull Seismol Soc Amer, 96(3): 757–795.

Zhang C, Cao X, Qu K. 1990. Focal Mechanism Solutions of Chinese Earthquakes. Beijing: Academic Periodical Publishing House. (in Chinese)

Zheng Y, Ma H S, Lv J, et al. Source mechanism of strong aftershocks ($M_S \geqslant 5.6$) of the 2008/05/12 Wenchuan earthquake and the implication for seismotectonics. Sci China Ser D-Earth Sci, 52(6):739–753.

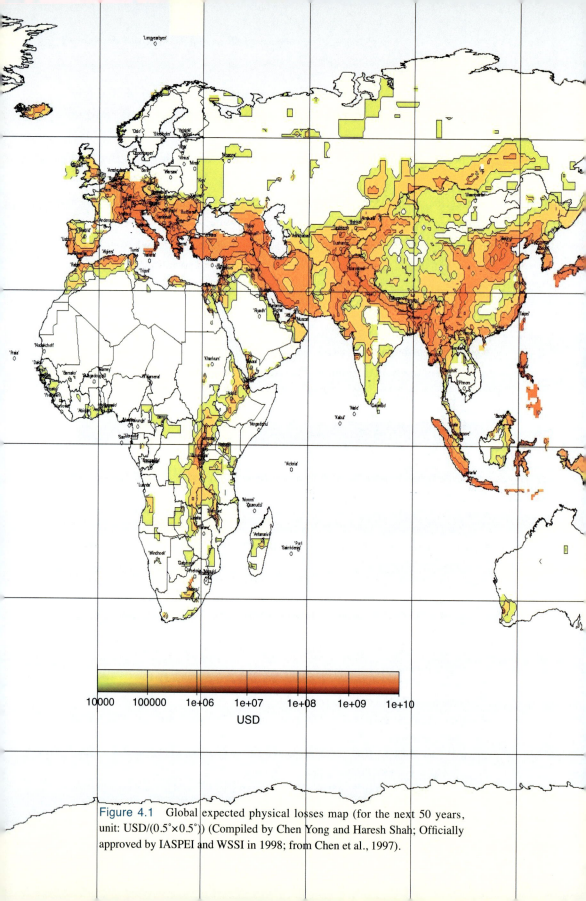

Figure 4.1 Global expected physical losses map (for the next 50 years, unit: USD/(0.5°×0.5°)) (Compiled by Chen Yong and Haresh Shah; Officially approved by IASPEI and WSSI in 1998; from Chen et al., 1997).

Seismic hazard and risk assessment

4

161 / Seismic hazard assessment
167 / Building code
172 / Risk management
180 / Did the reservoir impoundment trigger the Wenchuan earthquake?

The prediction program in China has provided useful experience and taught many lessons. The most important lesson is that, regardless of its future potential, it is presently impractical to rely on prediction to prevent earthquake disasters. The practical approach is to strengthen the resilience of our built environment based on an assessment of seismic hazard (Figure 4.1).

4.1 Seismic hazard assessment

A key question that must be addressed in earthquake disaster reduction is: how much loss might a city or region suffer in future earthquakes? Two components form the basic structure of a loss estimate study. The first component, seismic hazard analysis, involves the identification and quantitative description of probable strong ground motion caused by future earthquakes; the second component, seismic risk analysis, involves a vulnerability assessment of buildings and other man-made facilities regarding probable earthquake damage and the losses that may result from this damage. Seismic risk is determined by multiplying seismic hazard and vulnerability, so that reduction of high seismic hazard requires low vulnerability (e.g. strong construction).

Quantification of earthquake disaster is the basis for the reduction of natural disasters and risk management, in which seismic hazard analysis and seismic risk assessment are the key scientific problems (Figure 4.2).

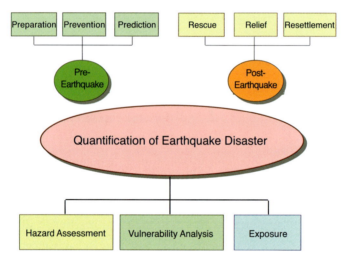

Figure 4.2 Quantification of earthquake disaster is an essential scientific basis for disaster reduction. Prevention: construction of defensive engineering works, land use planning and the formulation, dissemination and maintenance of evacuation plan. Preparedness: arrangements for emergency warnings to be issued and the effectiveness with which public officials can mobilize an evacuation plan. Reconstruction: long-lasting activities that attempt a return to "normal" after severe devastation. Such devastation can occur even in areas apparently well prepared for disaster.

Earthquake hazard is defined as the probability that a certain value of a macroscopic intensity or of a ground motion parameter (i.e. particle acceleration, velocity and displacement) will not be exceeded at a specific site in a specific period of time. Seismic hazard describes the potential for dangerous, earthquake-related natural phenomena such as ground shaking, fault rupture, or soil liquefaction. These phenomena can cause adverse consequences to society such as the destruction of buildings or loss of life.

The output of seismic hazard analysis can be a description of the intensity of shaking, or a map which shows the level of ground shaking in various parts of the country that has an specific chance of being exceeded.

Compared with the fairly recent upsurge of prediction studies, seismic hazard assessment in China had an earlier but very slow start.

1957—First intensity zoning map of China

Earthquake intensity is a scale which is used to calculate and compare the severity of earthquakes in terms of felt effects and damage. In China, the first earthquake intensity scale was developed in 1957—the "New scale of seismic intensity of China", with a corresponding zoning map. The twelve-level Chinese seismic intensity scale is similar to the Modified Mercalli Intensity (MMI) scale (Table 1.1).

Before 1957, earthquake hazard assessment was done only for critical engineering projects. In 1957, based on the analyses of historical earthquakes and geological structure, Lee (1957) published China's first earthquake intensity map (Figure 4.3) showing expected seismic intensity that would be produced by future earthquakes but with no temporal specifications, and the government published the expected seismic intensity for 297 municipalities. However, the government also introduced the policy that no seismic design was required for ordinary buildings in regions of intensity less than VIII and specified some limits on building height and construction style for regions of intensity IX or greater. In 1959 and 1964, two versions of Construction Standards for Earthquake-prone Areas were developed but not officially adopted. There was a lack of seismic design for most areas of China, blamed on the very poor economic conditions at the time (Chen et al., 2010).

1977 intensity zoning map of China

In 1977, a year after the great Tangshan earthquake of magnitude 7.8, SSB published the Seismic Hazard Map of China (1:3,000,000) (State Seismological Bureau, 1977), a fundamentally improved version from the 1957 map of Lee (1957). The area of intensity VI and above in this hazard map is 5.85 million km^2, and the area of intensity VII and above is 3.08 million km^2. This map was made based on a long-term earthquake forecast. The mapped basic intensity is defined as: in the specified area, over a specified time interval, in ordinary site conditions, this is the highest seismic

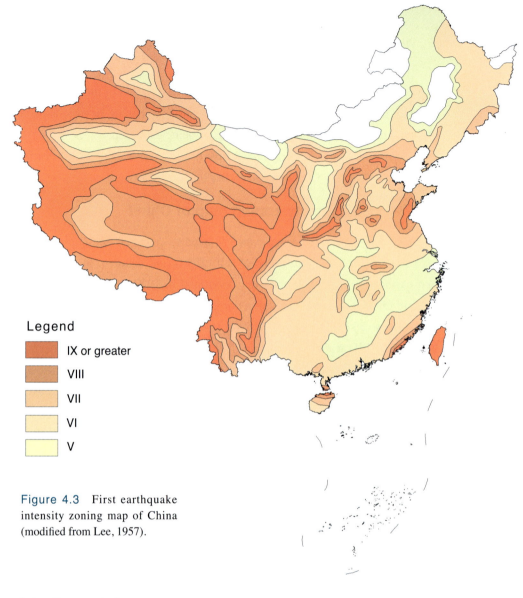

Figure 4.3 First earthquake intensity zoning map of China (modified from Lee, 1957).

intensity expected.

This hazard map incorporated macroscopic description and physical measures of PGA (Peak Ground Acceleration) and PGV (Peak Ground Velocity). The basic intensity was defined as the highest intensity that can be encountered in a region of ordinary site condition in next 50 years.

1990 intensity zoning map of China

In 1992, a third edition, the China Seismic Intensity Zoning Map (1:4,000,000) (1900) was promulgated. In this map the basic intensity was re-defined as the intensity that has a 10% probability of being exceeded in the next 50 years. The area of intensity VI and above in this hazard map is 7.59 million km^2, and the area of intensity VII and above is 3.97 million km^2. Compared to the 1977 zoning map, the areas of intensity VI and above, and VII and above, increased by 30%, the area of intensity VIII and above decreased by 7.5%, and the area of intensity IX and above decreased by 71%.

In the 1980's, the probabilistic method of seismic hazard analysis was introduced into China from the United States. Underestimation of high-risk areas, and overestimation of low risk areas is a common problem with probabilistic analysis. For example, the intensity of Wenchuan region is IX in the 1957 map, VIII in the 1977 map, and VII in the1990 map. Wenchuan was assigned to an earthquake zone of intensity VII. Therefore, little consideration had been given to earthquake hazard there.

The results obtained from hazard analysis of the 1990 map provided the basis for the seismic design code GBJ11-89 for seismic design of buildings (Department of Construction of China, 1989). Three levels of seismic protection were proposed, that is, to resist a "minor" earthquake without damage: in a 50 years period, the modal intensity has a probability of 63% of being exceeded; to resist a "moderate" earthquake with reparable damage: in a 50 years period, the occurrence of a "moderate" earthquake of basic intensity has a probability of 10% of being exceeded. The average difference between the basic intensity and the modal intensity is 1.55 degree; to resist a "major" earthquake in the next 50 years, the occurrence of a "major" earthquake has a probability of 1.5%–3.0% of being exceeded, where the intensity of a "major" earthquake is approximately one degree higher than the basic intensity.

2001—Latest hazard map of China

In order to reflect the latest achievements of seismic hazard analysis, and to transfer from earthquake intensity to dynamic seismic parameters, a new 1:4,000,000 "Zoning Map of China (GB18306-2001)" was promulgated in 2001 (Figure 4.4). This consists of two zoning maps. One is the PGA distribution map showing areas with 10% probability of exceeding PGA levels of < 0.05 g, $\geqslant 0.05$ g, $\geqslant 0.1$ g, $\geqslant 0.15$ g, $\geqslant 0.20$ g, $\geqslant 0.30$ g and $\geqslant 0.40$ g in the next 50 years. The other map is a response spectrum characteristic period zoning map.

As examples, the appropriate intensity and design PGA in the 2001 code for a few places affected by the 2008 Wenchuan earthquake are listed in Table 4.1 (Chen et al., 2010).

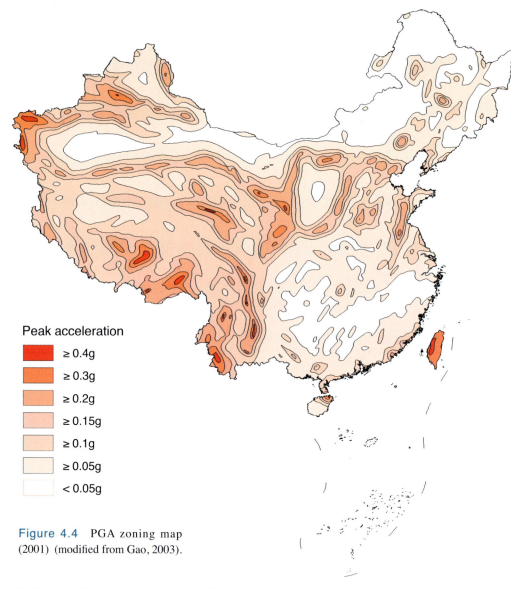

Figure 4.4 PGA zoning map (2001) (modified from Gao, 2003).

Table 4.1 Examples of seismic fortification intensity and design PGA for cities/counties in Sichuan and some other Chinese cities specified in the 2001 seismic design code (GB50011-2001)

Fortification intensity	Design PGA (g)[*]	Example cities/counties in Sichuan[**]	Example cities in other parts of China
≥ IX	≥ 0.40	Kangding (VI)	Taichung, Gulang, Dangxiong
VIII	0.30	Mianning (VI)	Longling, Haiyuan, Haikou
VIII	0.20	Songpan (VIII)	Tangshan, Beijing, Lhasa
VII	0.15	Pingwu (XI, VIII–0.2)	Haicheng, Hongkong, Tianjin
VII	0.10	Beichuan (XI, VIII–0.2)	Xingtai, Chengdu, Shanghai
VI	0.05	Mianyang (VIII, VII–0.1)	Changsha, Wuhan, Yanan

[*] Fortification intensities VII and VIII are each subdivided into two levels using the design PGA.
[**] The first Roman numeral in parentheses is maximum intensity of the Wenchuan earthquake of 12 May 2008 within the city/county territory. The second Roman numeral, if present, is the revised fortification intensity in the 2008 amendment to the 2001 code for the same location, followed by the revised design PGA.

4.2 Building code

Scientific aspects of hazard assessment provide the basis for risk mitigation such as accurate estimation of the largest earthquake that the Longmenshan fault zone can generate, its recurrence time, and the level of shaking it produces. The results of seismic hazard analysis provide the basis for anti-earthquake design, which is the main engineering measure for reduction of earthquake disasters.

The seismologists performing seismic hazard analysis are really carrying out one part of an engineering process or one part of the social disaster reduction process. The end product of their analysis is an expression of seismic hazard or threat that is oriented toward some specific use. This product may take the form of a simple, single-value characterization of earthquake ground motion such as MMI or PGA, or more complex multi-value characterizations, such as response spectra. In any case it can be readily used to achieve a practical objective.

The basic philosophy of a seismic building code is that when buildings are designed to be earthquake resistant, damage to buildings may be reduced, economic losses may be minimized, and the loss of human lives may be avoided. Under the action of minor earthquakes, which may occur frequently, the buildings will be undamaged, or only slightly damaged and will continue to be serviceable without repair. Under the action of an earthquake of the specified seismic-protection intensity, buildings may be damaged but will be serviceable after minor repair, or without repair. The seismic-protection intensity is that approved by the state authority to be used as a basis for the seismic protection of a region. In the majority of cases, the seismic-protection intensity is equal to the basic intensity, which was defined in the previous section. Under the action of a major earthquake or rarely occurring severe earthquake, the buildings will suffer irreparable damages but will neither collapse nor suffer failure that would endanger human life. This basic objective is in agreement with the codes of many countries. All provisions in the code, whether for the determination of the values used in seismic calculation, improvement of structural integrity, limiting the height or other dimensions of the structure, or for other purposes, are written to attain this basic objective.

In 1959 and in 1964 two building codes (59 code and 64 code) were developed, but not formally promulgated. China's first building code was promulgated in 1978.

Building code (TJ11-78)—effective 1979–1990

In December 1974, a Code for Seismic Design of Industrial and Civilian Buildings came into effect. In China, many codes covering different aspects of the quality and safety of built structures work together to have the effect of a (mandatory) national building code. The seismic design code is the most important of these codes and defines construction standards to prevent collapse or severe damage due to earthquake shaking. Other codes deal with foundations, sound insulation, lighting, plumbing, fire prevention, convenience of use, etc. Chinese seismic design makes a distinction between "basic intensity" and "design intensity". They have the same physical meaning and are basically a measure of the severity of ground shaking. The former represents the expected seismic shaking based on scientific analyses and a certain probability measure. The latter represents the degree of shaking that the buildings are built to withstand which takes into account economic and societal considerations. In addition, structures are classified into different categories such as critical infrastructure and regular buildings. The design intensity mentioned in this book is for regular buildings unless otherwise specified. For regular buildings, the 1974 code required seismic design in regions of basic intensity VII or greater, an improvement over the 1957 policy (VIII or greater), however it allowed the design intensity to be lower than the basic intensity by one level, although not less than VII. This means that for basic intensities IX, VIII, and VII, the design intensities could be VIII, VII, and VII, respectively. The reason given for allowing the design intensity to be lower than the basic intensity was the lack of financial resources.

The 1974 code performed miserably for two reasons. Firstly, the basic intensity, and consequently the design requirement, was too low for many regions. This was a reflection of the state of knowledge of the seismic hazard, which could be improved only through future scientific research. Secondly, no seismic design was required for regions of basic intensity VI (and lower). This was a fatal mistake in risk mitigation policy that could have been avoided. Like the 1966 Xingtai earthquake, both the 1975 Haicheng and 1976 Tangshan earthquakes occurred in regions of basic intensity VI where no seismic design was required. These three earthquakes caused over 91% of the building collapses due to earthquakes from 1949 to 1976. In the Haicheng region, pre-earthquake evacuation, the local construction style, and the time of the earthquake (7:36 pm) saved many lives (Wang et al., 2006). But in Tangshan, over 240,000 lives were lost under collapsed buildings (Chen et al., 1988).

The Code for Seismic Design of Industrial and Civilian Buildings (TJ11-78), effective 1979–1990, and referred to in China as the "1978 code" because of its completion date, set a slightly higher standard than the 1974 version for some building types. In May 1984, the predecessor of the Ministry of Construction implemented temporary provisions on seismic design for large cities in regions of basic intensity VI. This was a revolutionary amendment to the 1978 code which did not require seismic design for regions of basic intensity VI or less.

Seismic Codes were reviewed and amended many times as a result of severe damage during consecutive strong motion earthquakes. Since the 1980s, seismically resistant structure research has changed track from static analysis to dynamic analysis (Figure 4.5). Dynamic analysis, aided by rapid theoretical and computational advances, provides more precise design methods for structures subjected to strong motion earthquakes. However, despite advances in the theoretical modelling of structures, it is not yet possible to accurately predict the response of structures when subjected to future earthquakes, whose characteristics are unknown. It is expected that the seismic code will be improved as hazard analysis research progresses.

Figure 4.5 Simulation test on shaking table for improving the seismic design code of buildings (courtesy of Wang Lixin).

Building code (GBJ11-89)—effective 1991–2001

In 1989, based on the 1977 Seismic Hazard Map, the Earthquake Office of Ministry of Construction completed the Code for Seismic Design of Buildings (GBJ11-89), which is referred to as the 1989 code and was effective 1991–2001. In 1990, this third edition was based on the China seismic intensity zoning map (1990) described in section 4.1. The new code largely matched international standards and required seismic design for regions of basic intensity VI and above in all of China, not just the large cities.

Building code (GB50011-2001)

In 2001, the State Bureau of Quality and Technical Supervision published a new Seismic Ground Motion Parameter Zonation Map of China (1:4,000,000) (GB18306-2001). It consists of two maps, one is the seismic zoning map of PGA, and the other is the zoning map of the characteristic period of the response spectrum. In 2002 a new version of the seismic design code (GB50011-2001) based on this map, referred to as the 2001 code, came into effect. In both the 1989 and 2001 seismic design codes, the expected ground acceleration with 10% probability of being exceeded in 50 years is used to define the basic intensity. In both codes, the required construction standard is in terms of regional "fortification intensity" (a more general version of "design intensity" which was meant for individual structures) and "design Peak Ground Acceleration" ("design PGA"). There are also provisions regarding spectral response, soil type, structure category (such as critical vs. regular), and so on. Examples of the fortification intensity and design PGA in the 2001 code for a few places affected by the 2008 Wenchuan earthquake were listed in Table 4.1, and compared with some large cities in China.

Importance of complying with the seismic design code

The 1989 or 2001 seismic design codes regulated post-1990 construction in urban areas of Wenchuan. The two versions of the code specified similar standards for Sichuan. The fortification intensity (about VII) defined in the 2001 code for the Longmenshan fault zone area is now recognized to be too low. After independent surveys of building damage caused by the Wenchuan earthquake, various engineering groups reached the same conclusion: on average, buildings that incorporated the seismic design suffered much less than those that did not (Risk Management Solutions, 2008; Civil and Structural Groups of Tsinghua University et al. 2008).

The Wenchuan earthquake initiated a new era in earthquake risk mitigation in China (Figure 4.6). An amendment to the 2001 seismic design code is to raise the fortification intensity from VII to VIII for the Longmenshan fault zone area and some neighbouring regions. The rigor applied in enforcing the (amended) code when rebuilding in the disaster area has been unprecedented. For rebuilding, no distinction is made between urban and rural areas in enforcing the code, although compliance with the code in most other rural areas of China is still voluntary and will be guided by various "demonstration" projects for years to come. There are also very encouraging indications of risk mitigation awareness throughout the area where rebuilding is being done. For example, a popular slogan for commercial real estate promotion posters displayed on highways is "built to resist magnitude 8 (earthquakes)".

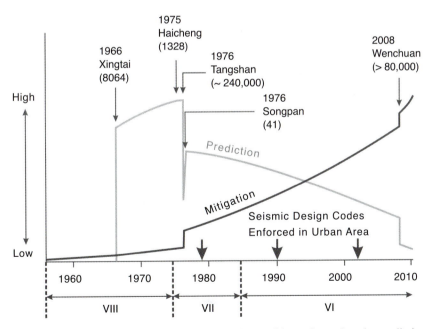

Figure 4.6 Schematic illustration of changes in confidence in earthquake prediction (gray) and emphasis on seismic risk mitigation (black) in China over the past five decades and relationship with major earthquakes. Death toll for each earthquake is shown in parentheses. Minimum expected seismic intensity for which seismic design was required by regulations or codes is shown below for different periods. Prior to 1957, seismic hazard was considered only for critical engineering projects (from Chen et al., 2010).

4.3 Risk management

Quantification of disaster

Quantification of disaster is a basis of risk management.

Most methods for estimating earthquake losses require a detailed inventory of the facilities and structures within the region in concern. In many parts of the world, however, the difficulty in building such a database and insufficient knowledge of local faults and soil conditions render impossible the undertaking of this kind of loss studies.

Chen et al. (1997) proposed a new approach: Quick and approximate estimation of earthquake loss based on macroscopic index of exposure and population distribution.

Chen's approach to exposure bypasses the data collection problems of the traditional method by employing a macroscopic indicator to represent the total exposure directly. This study expresses exposure according to the latter approach. It uses social wealth as the macroscopic indicator, and uses Gross Domestic Product (GDP) to represent social wealth and to estimate earthquake loss. The fundamental assumption of this holistic technique is that the number of man-made facilities is directly proportional to social wealth, specifically to GDP.

The vulnerability relation in the macroeconomic method can be determined from earthquake loss case studies. According to the income economy of a given area, the vulnerability relations (MDF—mean damage factor) can be divided into High income economy (HIE), Mid income economy (MIE), and Low income economy (LIE), In the new classification system (HIE, MIE and LIE) , the loss vs. intensity curves exhibit some differences, probably because of different earthquake resistant capabilities. The upper (LIE) line and lower (HIE) line of Figure represent the vulnerability curves of best and worst buildings in inventory method respectively (Figure 4.7).

The Map of world expected earthquake losses in next 50 years was showed in Figure 4.1 and in Table 4.2 by using macroeconomic method.

It can be seen from Table 4.2 the expected earthquake loss of mainland of China is about 5.4 billion US dollars (2000) per year. By using macroeconomic method, the loss estimation of Wenchuan earthquake was reported in the next day of mainshock: 200 billion yuan (about 35 billion dollars).

Many lessons in risk management were obtained from Wenchuan earthquake.

Table 4.2 20 countries or regions of the world with the highest expected loss in the next 50 years (exceeding probability 10%)[*]

Region	Population in 2009 (million)	GDP in 2009 (billion dollars)	Loss (billion dollars)	Loss per year (billion dollars)
World	6,500	40,800	4,000	80
Japan	150	5,600	708	14
China (Mainland)	1,320	5,475	270	5.4
United States	280	7,342	196	4.0
Indonesia	220	226	98	2.0
Italy	61	1,208	80	1.6
Greece	11	123	80	1.6
Mexico	102	335	58	1.2
Philippines	81	84	39	0.8
Turkey	71	181	38	0.8
Chile	17	52	38	0.8
Argentina	38	294	38	0.8
India	1,103	356	36	0.8
Colombia	36	64	32	0.6
Pakistan	155	65	32	0.6
Russia	162	440	32	0.6
Peru	25	61	30	0.6
Iran	63	111	22	0.4
Venezuela	25	67	18	0.4
Afghanistan	22	10	14	0.4

* Population and GDP data are from the report of World Bank.

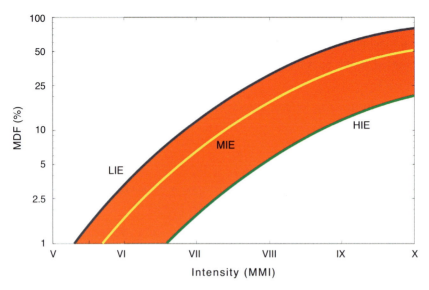

Figure 4.7 The vulnerability relation used in the macroeconomic method.

Fortification standards should be increased

The Wenchuan earthquake overwhelmingly demonstrated the vital importance of seismic risk mitigation. In contrast to Tangshan 1976 where seismic design was not required, seismic design had been required for the Wenchuan area since 1990. In 1976, most of the seismic intensity XI area of Tangshan was razed to the ground, but in 2008, many buildings in the intensity XI areas of Wenchuan did not collapse. The difference was due to improved construction standards. For regular buildings, there was no seismic design in Tangshan in 1976 for resisting any earthquake shaking, but three decades later limited seismic design was required for the Wenchuan area. Although the design standard in the Wenchuan area was later recognized to be too low, those buildings that did meet the standard suffered much less than those that did not. However, progress in practice lags very far behind progress in knowledge and regulations.

It is commonly accepted that the most effective way of minimizing the impact of earthquakes on human life and the economy is to strengthen our built environment, based on scientific assessment of earthquake hazard and risk.

An important lesson from the Wenchuan earthquake is that the fortification intensity of mainland China is too low. In most Wenchuan earthquake-hit areas, the actual intensity of shaking caused by the Wenchuan earthquake was IX or X, but the fortification intensity (from the intensity zoning map) was only VII! The fortification intensity of Longmenshan region had decreased from IX (1957 zoning map) to VII (2001 zoning map). In China, there are many examples of underestimating intensity in the past decade. The most urgent thing is to improve the hazard assessment method and raise the fortification standard for building structures.

Chen Zhun of China's Ministry of Construction said "about half of all houses in China need to be demolished or reconstructed in the next 15–20 years". According to his proposals, houses built before the 1977 building codes should be torn down and rebuilt. Therefore a few hundred million m^2 of new housing should be constructed every year, but what criteria should be used in building the new houses? Clearly they should not be built according to old standards. The Wenchuan earthquake indicates that new standards should be used for housing.

All engineers who have been to the Wenchuan disaster area, as well as the authors ourselves, can say that given the construction standards, engineering technology, and financial resources available prior to the earthquake, much more could have been done to minimize the disaster.

The significant improvement in seismic risk mitigation over the previous 32 years proved to be extremely valuable in the 2008 Wenchuan earthquake. Many relatively new buildings that had been built to the new seismic design specifications withstood violent shaking, saving numerous lives. But there were also some serious problems that had devastating consequences.

1. There was a major disconnection between the enactment and enforcement of mitigation regulations. Collapsed or severely damaged buildings that had been built recently, but not to standard, or earlier, but not retrofitted, were common in the disaster area.
2. Little attention was paid to potential secondary hazards such as the earthquake-induced landslides which were responsible for many deaths.
3. No seismic design was required in rural areas. Even today, the seismic design code is enforced only in urban areas. The Wenchuan earthquake caused much more extensive collapse of residential buildings in rural areas than in urban areas.

4. Community facility structures such as schools and medical centers were considered regular buildings and in practice often received less attention in terms of seismic design. On average, these buildings performed more poorly than residential buildings.

Some of these problems will be revisited in the following section.

Earthquake-safe Rural Housing Demonstrations

Rural residences, consisting mostly of one or two-storey masonry buildings, suffered much more extensive collapse than urban residences. The reason is that there was generally no seismic design for rural housing. Although the mandatory design code is not limited to urban areas, historically there has never been a mechanism to enforce it in the countryside. A survey conducted in 2003 in a few villages within the administrative areas of Mianzhu and Shifang cities of Sichuan Province showed that 98% of the houses inspected had no formal engineering design, and in 97% no protective measures against earthquakes had been taken (Zhang et al., 2009).

At the beginning of the 21st century, some scientists and government organizations have begun to make efforts to mitigate earthquake risk in the vast rural areas of China. Many handbooks, written in plain language to teach peasants the basic techniques for making their houses more resilient to earthquake shaking were published (Figure 4.8). Some local governments started projects involving Earthquake-safe Rural Housing Demonstrations. Under these projects, some peasants who could afford the basic cost were selected and supplied with design plans and subsidized by the government and corporate sponsors, and new houses were built for them that strictly complied with the seismic design code (Zhang et al., 2009). The purpose was to motivate other peasants to build their houses in a similar manner, as a remedy for the lack of enforcement of seismic design regulations. In 2005, the China Earthquake Administration (CEA) launched a similar project but on a larger scale, and in 2006 began to build these "demonstration" houses in selected areas. In 2007, the State Council began to promote this project across the country. Some rural communities along the Longmenshan fault zone benefited from these demonstration projects. The seismic design code specified a fortification intensity of VII for Yanjing Village in Qingping County in the suburb of Mianzhu City. In this area, the actual intensity of shaking caused by the Wenchuan earthquake was X, but houses built in 2004 to the code-specified standard under a demonstration project suffered only very minor damage. More remarkable is the excellent performance of some new houses in the same village built after 2004 but before the Wenchuan earthquake. With little subsidy, these peasants still decided to build their houses to the same standard as those under the 2004 demonstration project, and that decision paid them handsomely. For most other rural areas affected by the Wenchuan earthquake, these measures came too late.

In less common situations, disaster mitigation requires more than mere compliance with the seismic design code. For example, some destruction in the Wenchuan earthquake disaster area was not caused by shaking but by secondary hazards or surface fault breaks. The worst case in this respect is the town of Qushan, which was situated in a beautiful narrow valley between steep mountain slopes and was the location of the Beichuan County government. Of at least 12,000 fatalities in Qushan, about two-thirds are estimated to have been caused by landslides or rock avalanches induced by the earthquake. The massive Wangjiayan landslide overran several blocks of streets and took thousands of lives. A high-quality six-storey county government office building, built in 2003, missed being hit by the landslide, but suffered total collapse. The Wangjiayan landslide and the active fault that runs through the town were hazards which had long been mapped by geologists. The lack of adequate consideration of earthquake hazard in city planning was a tragic failure of risk mitigation.

Figure 4.8   Rural housing and house plan (Henan Province, 2009).

Figure 4.9 The 156-meter-high Zipingpu Dam (fortification intensity VII+), 20km east of the Wenchuan earthquake epicenter, suffered intensity X vibration during the Wenchuan earthquake. After inspection, engineers determined that Zipingpu dam suffered cracking but was structurally sound (courtesy of Gong Jianya).

4.4 Did the reservoir impoundment trigger the Wenchuan earthquake?

The 16 January 2009 issue of *Science* magazine published, in its "News of the Week" section, an article entitled "A Human Trigger for the Great Quake of Sichuan?" (Kerr et al., 2009). The article boldly proposed "Zipingpu reservoir may have triggered the Wenchuan earthquake". Similar reports also appeared in the media six months after the Wenchuan earthquake, such as in the Daily Telegraph website in UK (Moore, 2009) and the New York Times (Lafraniere, 2009).

The Zipingpu dam and reservoir is a large-scale water control project situated on the Minjiang River in Sichuan Province (Figure 4.9). The dam was constructed in March 2001, water impoundment started in September of 2005, and the project was completed in 2006. As one of the few high dams in China, this rock-fill dam with a reinforced concrete face has a maximum height of 156 meters and a storage capacity of 1.112 billion m^3. The nearby Zipingpu Hydropower Plant is located 20 km east of the epicenter and it was damaged by the earthquake. A recent inspection indicated that damage to the dam was less severe than initially feared, and it remains structurally stable and safe.

Reservoir-triggered seismicity, or reservoir earthquakes, refer to substantially enhanced seismicity (both in frequency and magnitude) near a reservoir after impoundment. The first reported reservoir-triggered earthquake occurred in 1931 near Marathon reservoir in Greece. Since then, it has been realized that activities such as water impoundment or dam construction can trigger earthquakes. However, most reservoir earthquakes are moderate or weak events, which present no great damage to the dams. Only 18 reservoir dams have suffered damage from reservoir-triggered earthquakes, the strongest of which registered a magnitude of M6.4.

Statistics on about one hundred cases of reservoir-triggered earthquakes show that these events share the following characteristics:
1. Reservoir earthquake epicenters are close to the reservoir, mostly within 5 km. Their focal depths are mostly shallower than 5 km and very few are deeper than 10 km. The distance between the reservoir

and the hypocenter (rupture initiation point) of the Wenchuan earthquake, and ruptured faults, is over 20 km, much greater than 5 km. Zipingpu reservoir is located in a seismically active region, and at least five moderate or strong historical earthquakes had occurred nearby. A digital seismic network was installed to monitor seismicity during construction of the reservoir, and seismic observation started on 16 August 2004, prior to water impoundment. Statistical studies of earthquake catalogs provided by the Reservoir Earthquake Institute of Sichuan Earthquake Administration (Zhang et al., 2009), showed that the frequency and magnitude of earthquakes near Zipingpu reservoir did not change substantially since impoundment, although the water level changed. There is no evidence for any substantial increase in seismicity after impoundment.

Table 4.3 Major earthquakes triggered by reservoirs in the world (Chen, 2009)

Reservoir name	Height of dam (m)	Storage capacity ($10^8 m^3$)	Time of impoundment initiation	Time of starting seismicity	Time of the strongest earthquake	Magnitude M_S
Koyna, India	103	27.8	1962-06	1963-10	1967-12	6.4
Xinfengjiang, China	105	115	1959-10	1959-11	1962-03	6.1
Kinnersani, India	61.8		1965	1965	1969-04	5.3
Chirkey, USSR	233	27.8	1974-07		1974-12	5.1
Marathon, Greece	63	0.4	1929-10	1931	1938	5.0
Kremasta, Greece	165	47.5	1965-07	1965-12	1966-02	6.2
Monteynard, France	155	2.75	1962-04	1963-04	1963-04	5.0
Tongjiezi, China	74	3	1992-04	1992-04	1994-12	5.5
Bajina Basta, Yugoslavia	89	3.4	1967-06	1967-07	1967-07	5.0
Kanba, Zambia	123	1750	1958-12	1959-06	1963-09	6.1
Aswan, Egypt	111	1640	1968		1981-11	5.6
Oriville, USA	235	4.4	1967-11		1975-08	5.5
Volia Grande, Brazil	56	23	1973		1973	5.0

2. Reservoir earthquakes are usually weak. The strongest reservoir-triggered earthquake is an $M6.4$ event, which occurred in December 1967 at Koyna reservoir, India, located on the stable Deccan Plateau. Its epicenter is 3 km to the south of the dam. The dam is 103 m high, and impoundment began in 1962, after which about 450 earthquakes occurred within three years. On a global scale, most reservoir earthquakes are weaker than $M5$, and more than 80% of them are in the range of weak or micro earthquakes. There are 10 reservoir earthquakes with magnitude between $M5.0$ and $M5.9$, and only 4 events between $M6.0$ and $M6.4$. These are the $M6.4$ Koyna reservoir event, $M6.2$ Kremasta earthquake in Greece, $M6.1$ Xinfengjiang event in China, and the $M6.1$ Kariba event near the border of Zambia and Zimbabwe (see Table 4.3). In contrast, the Wenchuan earthquake has a magnitude of $M8.0$, much more than the historically strongest reservoir earthquake.

3. Very few reservoirs trigger earthquakes. There are more than ten thousand large-scale or medium-scale dams registered in the World Commission on Dams, and only 101 of them report reservoir-triggered earthquakes, only about 1% of the total.

4. Reservoir earthquake sequences generally follow a pattern of foreshock–mainshock–aftershock. Usually, reservoir earthquakes occur soon after impoundment as weak earthquakes (foreshocks). More frequent and stronger earthquakes follow until the strongest event (mainshock) occurs near the time when the water level peaks for the first time (Table 4.3). After the mainshock, earthquakes become weaker and less frequent (aftershocks). All reservoir seismicity follows this pattern without exception. In contrast to reservoir earthquakes, most tectonic earthquakes do not follow the foreshock–mainshock–aftershock pattern. Global seismic statistics demonstrate that less than 10% of major tectonic earthquakes have foreshocks. The 2008 Wenchuan earthquake did not occur in the early stage of impoundment of Zipingpu, nor did it occur when the water level was at peak height (in fact, it occurred when the water level was low). There were no foreshocks for the Wenchuan earthquake, and thus the earthquake sequence is not a foreshock–mainshock–aftershock one.

Therefore the Wenchuan earthquake is quite different to previous reservoir earthquakes, at least phenomenologically, since it does not show reservoir earthquake characteristics as far as magnitude, spatial distribution, and earthquake sequences are concerned.

5. The Wenchuan earthquake is caused by thrust faulting, which was never reported for previous reservoir earthquakes. The reservoir water acts as a load on the ground, which produces tensional stress beneath the ground and would facilitate strike slip faulting or normal faulting. That is why most reservoir earthquakes have strike-slip or normal faulting mechanisms. In contrast, the Wenchuan earthquake ruptured with thrust slip, resulted from strong horizontal pressure (not tension) that pushes hanging wall upward and footwall downward. Thrust faults are very common at convergent oceanic plate boundaries, but far less common in continental regions. The Wenchuan earthquake occurred on thrust faults in a way similar to subduction events at oceanic trench, suggesting its uniqueness and arguing for its significance for studies of continental dynamics.

Xu et al. (2006) reviewed the previously derived mechanisms of all major reservoir earthquakes in the world, and found that most of them are strike-slip or normal events, with strike-slip events slightly more prevalent than normal ones. No reservoir mainshock shows a thrust mechanism, though some aftershocks do. For example, Xu et al. (1996) reported that the 1967 M6.3 Koyna main shock was a strike-slip event, and some M4 aftershocks showed thrust mechanism. In 1977, there were also some thrust events near the Monticello reservoir (a medium to small scale reservoir, with capacity of 0.4 billion m^3) in South Carolina. But the earthquakes are weak with magnitudes no greater than M2.8, which makes it difficult to distinguish them from background earthquakes. Except for the above two regions, there are no reports of thrust earthquakes (particularly mainshocks) triggered by reservoirs worldwide.

The causal link between reservoir impoundment and triggered seismicity certainly remains a challenging problem and demands detailed studies, despite many published researches. We have only discussed the special case of the Wenchuan earthquake and the purported role of the

Zipingpu reservoir in triggering it, instead of expounding on a detailed analysis of comprehensive links between reservoir impoundment and triggered seismicity. Our preliminary conclusion is that the Wenchuan earthquake is very different from reservoir earthquakes from aspects of phenomenology and source mechanics, and therefore not triggered by the Zipingpu reservoir.

References

Burchfiel B C, Royden L H, Van Der Hilst R D, et al. A geological and geophysical context for the Wenchuan earthquake of 12 May 2008, Sichuan, People's Republic of China. GSA Today, 18(7): 4–11.

Chen Q F, Chen Y, Liu J, et al. 1997. Quick and approximate estimation of earthquake loss based on macroscopic index of exposure and population distribution. Nat Hazards, 15(2/3):215–229.

Chen Y. 2009. Did the reservoir impoundment trigger the Wenchuan earthquake. Sci China Ser D-Earth Sci, 52(4): 431–433.

Chen Y, Tsoi K L, Chen F B, et al. 1988. The Great Tangshan Earthquake of 1976: An Anatomy of Disaster. Oxford: Pergamon Press.

Chen Y, Chen Q F, Liu J, et al. 2002. Seismic Hazard and Risk Analysis: A Simplified Approach. Beijing: Science Press.

Civil and Structural Groups of Tsinghua University, Southwest Jiaotong University, Beijing Jiaotong University. 2008. Analysis on seismic damage of buildings in the Wenchuan earthquake. J Build Struct, 29(4): 1–9. (in Chinese)

Earthquake Office of Ministry of Construction.1990. Training Materials for Code for Seismic Design of Buildings GBJ11-89. Beijing: Seismological Press. (in Chinese)

Gao M T. 2003. New national seismic zoning map of China. Acta Seismol Sin,16(6): 639–645.

Henan Province. 2009. Rural housing and housing structure map set. http://www.zhufang123.com/. Accessed 15 Apr 2011.

Hu Y, Gao M, Xu Z, et al. 2001. Seismic Ground Motion Parameter Zonation Map of China GB18306-2001. Beijing: China Standards Publishing House.(in Chinese)

Huang Q H. 2008. Seismicity changes prior to the M_S8.0 Wenchuan earthquake in Sichuan, China. Geophys Res Lett. 35, L23308. doi:10.1029/2008GL036270

Chen J, Hayes G. 2008. Finite fault model—preliminary result of the May 12, 2008 M_W 7.9 eastern Sichuan, China earthquake. http://earthquake.usgs.gov/earthquakes/eqinthenews/2008/us2008ryan/finite_fault.php. Accessed 23 Mar 2011.

Kerr R A, Stone R. 2009. A human trigger for the great quake of Sichuan? Science, 323(5912): 322.

Lafraniere S. 2009. Possible link between dam and China quake. http://www.nytimes.com/2009/02/06/world/asia/06quake.html. Accessed 15 Apr 2011.

Lee S P. 1957. The map of seismicity of China. Chin J Geophys. 6(2):127–158. (in Chinese)

Lu M. 2006. Guidelines for Seismic Design in Rural Housing. Beijing:Seismological Press. (in Chinese)

Ministry of Construction of the People's Republic of China.1979. TJ11-78, Seismic design of industrial and civil buildings. Beijing: China architecture & Building Press.(in Chinese)

Ministry of Construction of the People's Republic of China.2001.GB50011-2001, Code for Seismic Design of Buildings. Beijing: China architecture & Building Press.(in Chinese)

Moore M. 2009. Chinese earthquake may have been man-made, say scientists. http://www.telegraph.co.uk/news/worldnews/asia/china/4434400/Chinese-earthquake-may-have-been-man-made-say-scientists.html. Accessed 15 Apr 2011.

Risk Management Solutions. 2008. The 2008 Wenchuan Earthquake: Risk Management Lessons and Implications. http://www.rms.com/Publications/2008_wenchuan_earthquake.pdf. Accessed 15 Apr 2011.

State Seismological Bureau. 1977. Seismic Hazard Map of China and Explanations. Beijing: Seismological Press. (in Chinese)

State Seismological Bureau. 1992. Explanations of Seismic Hazard Map of China (1:4000000). Beijing: Seismological Press. (in Chinese)

State Seismological Bureau. 1996. Introduction to Seismic Hazard Map of China (1990). Beijing: Seismological Press. (in Chinese)

Tian W, Chen Z, Xiang X, et al. 2008. Analysis of and reflections on building damage caused by the 12 May Wenchuan earthquake in Sichuan. Sichuan Constr Sci, 34(6): 123–129. (in Chinese)

Wang K L, Chen Q F, Sun S H, et al. 2006. Predicting the 1975 Haicheng earthquake. Bull Seismol Soc Amer, 96(3): 757–795.

Wang L, Tao Y. Yuan Y, et al. 2005. Overview of the Earthquake-safe Rural Housing Demonstration Project of China. Northwest Seismo J, 27(4):305–311. (in Chinese)

Wang W M, Zhao L F, Li J, et al. 2008. Rupture process of the M_s 8.0 Wenchuan earthquake of Sichuan, China. Chin J Geophys, 51(5): 1403–1410. (in Chinese)

Xu G Y, et al. 2006. Proceedings on Reservoir Triggered Earthquakes in the World. Kunming: Yunnan Science and Technology Press. (in Chinese)

Zhang H, Xiang D, Hu Y. 2009. Earthquake-safe Rural Housing Project in Deyang City passed the test of the great Wenchuan earthquake. Cities and Hazard Mitigation, 3: 25–28. (in Chinese)

Zhang Z W, Cheng W Z,Zhang Y J, et al. 2009. Research on seismicity and source parameters of small earthquake in Zipingpu Dam before Wenchuan M8.0 earthquake. Earthquake Res China, 25(4): 367–376. (in Chinese)

Emergency response and rescue 5

- 188 / People-oriented rescue principle
- 199 / Open and transparent relief information
- 206 / The breadth and diversity of donors
- 213 / Voluntary contributions

In this chapter, from the point of view of Human Sciences we will discuss: what has changed when the response to the Wenchuan earthquake is compared with previous earthquake disaster relief, and what are the new issues concerning natural disasters in China?

The Wenchuan earthquake rescue work had many new features, when compared to rescue work in previous earthquakes:

People-oriented rescue principle

Priority is given to rescue the wounded as soon as possible and ensuring the safety of people in disaster areas, as time means life (Figure 5.1).

Humane care and respect is given to not only the living, but also in respect of the victims.

Figure 5.1 Rescue in mountain regions. National earthquake disaster emergency rescue team squad of 40 arrived in Yingxiu, the hardest hit area of Wenchuan earthquake.

Open and transparent information provision

Information on relief operations after the Wenchuan earthquake was provided openly to the public and to the world, so that relief became an international concern.

China allowed foreign media access to the disaster scene, together with foreign professional rescue teams and medical teams who entered Wenchuan to help carry out rescue and relief operations.

The breadth and diversity of donors

Donations came from foreign governments, international organizations, foreign friends, Chinese overseas, and particularly from the broad Chinese public. Donations are vital not only for providing funds for emergency supplies, but for simulating and enabling the mobilization of relief groups. Up to 12 February 2009, the total donation from home and abroad was 41,742 million yuan (about 5.1 billion dollars).

Broad participation of volunteers

The scale of volunteer development showed an increasing maturity of society. From across the country came more than 20 million volunteers from all parts of Sichuan (including 40% of the students at university), providing a voluntary response in their own interpretation of the societal principle of "dedication, friendship, mutual aid, and progress".

Although the earthquake caused a huge disaster, it provided the impetus and opportunity to consolidate and improve the rescue system, emergency response and fund-raising mechanisms, as well as mobilizing and enhancing Chinese society. Much valuable experience and many useful insights were gained, which promise a brighter future in respect of China's relief and social development programmes.

5.1 People-oriented rescue principle

Immediate top level government response

Immediately after the earthquake occurred, President Hu Jintao announced that the disaster response would be rapid, saying, "disaster gives the order, time is life".

Just 90 minutes after the earthquake, Premier Wenjiabao, who has an academic background in geology, flew to the earthquake area to oversee the rescue work. He was the first official to reach the disaster area, 5 hours after the mainshock. The National Disaster Relief Commission initiated its "Level II emergency contingency plan", which covers the most serious natural disasters. The plan rose to Level I at 22:15 on 12 May. A national earthquake relief headquarters was established on 12 May, with Premier Wen Jiabao as its director.

A national earthquake emergency relief team of 184 people (consisting of 12 people from the China Earthquake Administration, 150 from the Beijing Military Area Command, and 22 from the Armed Police General Hospital) left Beijing in two military transport planes to travel to the disaster area , arriving 10 hours after the mainshock (Figure 5.2). In view of the distance of 1500 km between Beijing and Wenchuan, it may have been the fastest response in global natural disaster rescue history. Overnight the national earthquake disaster rescue team rushed to Sichuan, the worst hit area. In the emergency relief operations, which lasted 18 days, they successfully rescued 49 survivors, and removed the bodies of 1080 victims.

Figure 5.2 The national earthquake disaster rescue team of 184 people left Beijing in two military transport planes to travel to the disaster area, 1500 km away from Beijing, and arrived in Chengdu 10 hours after the earthquake, probably a record in the history of earthquake disaster rescue!

At the same time in Sichuan, Chongqing, Jiangsu, Hainan, and Liaoning, 19 provincial-level earthquake rescue teams amounting to more than 4000 people rushed to the scene to implement a rescue operation. Rescue teams at all levels rescued 371 survivors (China Earthquake Administration, 2008).

On the same day, China's Chengdu Military Area Command dispatched 50,000 troops and armed police to help with disaster relief work in Wenchuan County. The objective was to rescue the wounded as soon as possible, as well as ensuring people's safety and security, as part of the "people-oriented" principle of respect, a standard for social progress and performance.

So long as there was hope, they would never give up

At the beginning of the rescue effort, it was not possible to travel over roads that were completely damaged or to reach places that were blocked off by landslides. Landslides continuously threatened the progress of the search and rescue effort (Figure 5.3). Persistent rain, as well as rock slides and a layer of mud coating the main roads, hindered rescue officials' efforts to enter the disaster region. Falling debris also hindered rescue workers' progress as they attempted to cross the mountains.

The extreme terrain conditions precluded the use of helicopter evacuation, and over 300 Tibetan villagers were stranded in their destroyed village for five days without food and water before the rescue group finally arrived to help the injured and stranded villagers down from the mountain. A group of 80 men, each carrying about 40 kg of relief supplies, from a motorized infantry brigade under commander Yang Wenyao, tried to reach the ethnically Tibetan village of Sier at a height of 4000 m above sea level in Pingwu County.

By 15 May, China's Premier Wen Jiabao had ordered the deployment of an additional 90 helicopters, of which 60 were to be provided by the air force and army, and 30 were to be provided by the civil aviation industry. This brought the total of number of aircraft deployed in relief operations by the air force, army, and civil aviation to over 150, resulting in China's largest ever non-combat airlifting operation. On 31 May, a rescue helicopter carrying earthquake survivors and crew members crashed in fog and wind turbulence in Wenchuan County. No-one survived.

Over dozens of days and nights, domestic and foreign rescue teams

Figure 5.3 Mountain rescue was extremely difficult, including transport of the injured (photo by Ng Han Guan, AP).

searched for survivors. As long as there was a glimmer of hope, they continued the search for people trapped in the rubble.

A time limit of about three days is usually given for the survival of people trapped under the ruins; 3 days, therefore has been called the "golden 72 hours of rescue time." Three days after the earthquake, the military and civilian rescue workers did not give up hope, particularly in hard-hit Beichuan, and conducted an intensive search to find signs of life. A number of rescued people became small miracles in disaster survival and tenacity, as illustrated by the following examples.

196 hours: Wang Youqiong

At 18:45 on 20 May, in a village in Pengzhou County, an Air Force rescue team rescued Wang Youqiong, who had been trapped for 196 hours. This 60-year-old woman is the longest buried survivor of an earthquake who has been rescued, a miracle of life. It is reported that when the earthquake occurred, Wang was crushed by a beam, and trapped in the mountains. When People's Liberation Army soldiers discovered her, she was conscious.

179 hours: Ma Yuanjiang

At 0:50 on 20 May 31-year-old power plant worker Ma Yuanjiang was successfully rescued by Shanghai Fire Department rescue workers from the ruins of Yingxiu power plant buildings.

164 hours: Wang Huazhen

Yunnan Fire Department in Deyang rescued 50-year-old Wang Huazhen, in Hanwang Township of Deyang City. When rescuers found her, she was

Figure 5.4　164 hours after Wenchuan earthquake, Yunnan Fire Department in Deyang rescued a 50-year-old survivor, in Hanwang Township of Deyang City. This is a miracle of survival.

crying for help, weak but still alive (Figure 5.4).

150 hours: Yu Jinhua

Yingxiu Hydro staff member Yu Jinhua was successfully rescued at 8:10

pm on 18 May by Shandong Public Security Fire Department, having been buried in rubble for 150 hours. On-site amputation was necessary to free her, and the rescue took 56 hours, the longest time involved in taking a survivor out of danger.

148 hours: Shen Peiyun

On 18 May, the China national earthquake rescue team successfully rescued a 53-year-old man, Shen Peiyun, in the town of Yingxiu who had been trapped for 148 hours, after 8 hours of painstaking rescue work. After examination, Shen Peiyun's vital functions were found to be normal.

140 hours: Liu Xiaosheng

An 82-year-old survivor was Liu Xiaosheng, suffering from hemiplegia, paralysed and confined to bed in a house in Beichuan County. After the earthquake, since nearby buildings had collapsed with few survivors, and he could only lie in bed waiting for rescue. After140 hours he was rescued by the efforts of the Jilin Province fire brigade, and 20 search and rescue crew. After treatment, he was out of danger.

127 hours: Wu Jianping

In Dujiangyan, the Russian rescue team rescued a 61-year-old woman; the rescue took 16 minutes and involved taking her out from between two collapsed floors. This was the first of the international rescue operations to successfully rescue a survivor.

123 hours: Jiang Yuhang

After a 9 hour effort, Shanghai Fire Department rescue team rescued 20 year old Chiang Yuhang in Wenchuan Yingxiu, at the earthquake epicenter.

114 hours: German Bo Geda tourists

At 8:10 on 17 May, the Military rescue team in Wenchuan County successfully rescued the German Bo Geda tourists buried under the rubble for five days. In the same day, the rescue team of Guizhou Province successfully rescued another Chinese tourist in hardest hit area (Figure 5.5).

Figure 5.5 Rescue team of Guizhou Province in hardest hit areas (photo by Li Mingfang, Xinhua News Agency).

Respect for life

In order to show their deep condolences to Wenchuan quake victims of all nationalities, the State Council decided that 19 to 21 May 2008 would be national days of mourning. Meantime, national and overseas institutions had lowered their flags to half-mast in mourning, public recreational activities ceased, and the Foreign Ministry, Chinese embassies and consulates set out books of condolence. At 2:28 pm on 19 May, the entire Chinese nation observed three minutes silence (Figure 5.6).

Humane care was shown not only for the living, but also in respect of the lives of the victims.

Nearly 80,000 people died, therefore how to handle the remains, was not only a health issue, but also a cultural issue. To show respect for the living to the dead, comfort the hearts of the living, the Government from the humanitarian point of view to determine the identity of the deceased, cremation arranged by the Home Department, the conditions do not have the cremation, burial treatment, can not determine the identity of remains numbered, record, camera, extract DNA (Figure 5.7). These practices reflect respect for life. Better reflect the people-centered philosophy is the national day of mourning, mourning for three days, three minutes of silence, is a glorious shining humanity and dignity of citizens demonstrated.

The PRC State Council has designated each 12 May from 2009 as the national "Disaster Prevention and Mitigation Day", with the following message. "The people of China will always remember 12 May 2008 for its 'national calamity' (Figure 5.8). Too many disasters are engraved on the history of the Chinese nation, never to be ranked with our national glories.

Figure 5.6 National mourning for the 2008 Sichuan earthquake victims—Tiananmen Square, 19 May 2008, 2:10 pm, before the moment of silence.

Figure 5.7 May, at Mianyang City funeral home. Its staff number the ashes of victims of the earthquake, waiting for relatives of the deceased to claim. Before the earthquake, a daily average of 20 bodies was cremated; a day after the earthquake, the average was more than 100. Seven crematoria operated almost around the clock. Before cremation, forensic field photographs were taken, as well biological samples. These were numbered and filed; after cremation, ashes were placed in a red bag, each labelled with the appropriate file number (photo by Liu Jin, AFP).

Think of the disaster, then let us bury our grief, clear the ruins which were once homes, rebuild them, be bold, and live good lives." (Figure 5.9)

Figure 5.8 The National Disaster Day logo. A rainbow as an umbrella, forms the basic composition. The rainbow after rain signifies hope and a sunny future; the umbrella is a familiar symbol of protection and care. The two people represent joint work for disaster prevention and mitigation. The logo represents positive thinking and protection of people's lives and property.

Figure 5.9 The first anniversary of the earthquake. Students in the country mourn the victims in various ways.

5.2 Open and transparent relief information

Timely, open and transparent information

In order to create a situation of social stability in the face of major disasters, past governments at all levels had been accustomed to "singing a song", and not telling the public the whole truth. This situation completely changed when the Wenchuan earthquake occurred.

The sudden disaster of the Wenchuan earthquake caused a massive upset to the lives of millions of people, but it did not cause social panic, for the very important reason that there was a timely and full disclosure of information to the public. In the face of great disaster, the Government and the media had learned a lesson and learned to race with, and defeat, the rumors. The Government responded quickly and effectively, and the media reported full, timely, transparent and comprehensive information, in complete contrast to the Tangshan earthquake in 1976 where there was a news blackout.

Less than 10 minutes after the earthquake, the China Earthquake Administration (CEA) linked with the Xinhua News Agency to release the news to the community; following its prompt release, the information was quickly transmitted countrywide. Central Television was the first to interrupt the broadcast of other programs, and immediately start showing live coverage of earthquake disaster relief live. The next day, "People's Daily" and other newspapers started providing multi-page, multi-level reporting on the disaster and the relief efforts. Major web sites gave updates on the disaster and up-to-date information on the relief operations. However the public felt sorry that a big disaster had to occur before disaster information became timely, open and transparent so that the disaster could be overcome in a more self-confident, calm and unanimous state of mind.

On 8 August 2005, the Ministry of Civil Affairs and the State Secrecy Bureau jointly issued a notice declassifying information on the consequences of natural disasters, the total number of deaths and related information. This meant that the death toll in a natural disaster is no longer a state secret. The response to the Wenchuan earthquake shows that adoption of this principle will effectively reverse the largely passive role which the public had adopted to natural disaster and associated relief work in the past, as well as respecting the citizens' right to know (Figure 5.10).

On 8 January 2006, the State Council issued "the overall national public emergency contingency plan" making it a crime to delay, falsify, conceal or omit information on an important public emergency, or dereliction of duty, misconduct, disciplinary action in accordance with the law relating to the responsibility of staff. On 30 August 2007 the "Act to deal with unexpected incidents", was promulgated and then implemented on 1 November, which further standardized the responsibility of the Government. "Government information open regulations", which were promulgated in April 2007, and implemented on 1 May 2008, further clarified that the Government should be timely and accurate in its disclosure of information. China's information disclosure system has been set up, and citizens have the right to get information on a legal basis. During disasters reported in recent years, changes have occurred, especially after the Wenchuan earthquake, which was reported in full and on time. The party and the government have continuously increased their ability to take a proactive role in managing the future, further promoting a message of openness, and reform and full disaster reporting.

The Wenchuan earthquake has been of historical significance for the principle of information disclosure in China, as well as the enhancement of public trust in the media. We trust that this is not a one-off case, and

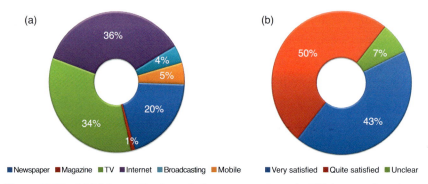

Figure 5.10 (a) Major media channels (newspapers, journals, TV, internet, broadcast, mobile) through which People understand Wenchuan earthquake's information. (b) Degree of satisfaction of people to the Media reports on the earthquake relief (Very satisfied, Quite satisfied, unclear). Source: Tsinghua University media research and investigation laboratory.

the experience will generate a healthy and long-lasting development of mechanisms for immediate open reporting of information.

International relief

A country's openness is measured by its readiness to accept assistance from the international community after disaster. The Chinese government is willing to accept any form of assistance from the international community, in support of the principle that the international community should support relief operations after a major disaster in any country.

China stated it would gratefully accept international help to cope with the earthquake. Many foreign reporters noted that China reacted to the disaster "rapidly and with uncharacteristic openness". Rescue efforts performed by the Chinese government were praised by the critical western media, especially in comparison with previous performance during the 1976 Tangshan earthquake. This was the first time that the Chinese media had clearly lived up to international standards.

China quickly allowed foreign media access to disaster scene, involving over 150 reporters from more than 20 countries. Soon after the earthquake, professional rescue teams from Japan, Russia, and R.O. Korea entered Wenchuan to search for survivors. The United Kingdom, Italy, France, Cuba and other nations, as well as the Red Cross, sent nine medical teams, so that 223 foreign medical personnel provided care for the wounded into the Wenchuan.

Japan sent the first foreign earthquake rescue team to arrive in China, the first international rescue team to enter China since the founding of New China in 1949 (Figure 5.11). The action of the Japanese government was the quick response of a good neighbour, and fully embodied the international spirit of unity, mutual help, and friendliness in countering disaster.

On 16 May, rescue groups from Japan, Russia, Singapore, R.O. Korea arrived to join the rescue effort (Figure 5.12). The United States shared its satellite images of the quake-stricken areas with Chinese authorities. The US also sent into China two U.S. Air Force C-17's carrying supplies, including tents and generators. 135,000 Chinese troops and medics were involved in the rescue effort across 58 counties and cities.

Many rescue teams, including that of the Taipei Fire Department from Taiwan, were reported ready to join the rescue effort in Sichuan as early as Wednesday 14 May.

Figure 5.11 Japan sent the first foreign earthquake rescue team to arrive in China (arrived 16 May).

Figure 5.12   Members of the Russian rescue team.

Rescue efforts

The emergency response and rescue work following the Wenchuan earthquake was performed in accordance with the "Law of the People's Republic of China on Protecting Against and Mitigating Earthquake Disasters" (1997), and was declared to be a successful test of the law.

Persistent heavy rain and landslides in Wenchuan County and the nearby area seriously affected rescue efforts. At the start of rescue operations on 12 May, 20 helicopters were deployed for the delivery of food, water, tents and emergency aid (Figures 5.13, 5.14), and also the evacuation of the injured and reconnaissance of quake-stricken areas. By 13 May, a total of over 15,600 troops and militia reservists from the Chengdu Military Region

Figure 5.13 An aerial view of tents housing displaced earthquake survivors in Sichuan Province on 24 May 2008 (photo by Evan Schneider /UN).

Figure 5.14 Relief materials are dropped over the earthquake-hit Qingchuan County, Sichuan Province 24 May 2008 (photo by Stringer, Reuters).

had joined the rescue force in the hardest hit areas.Wenchuan, that around 3000 survivors were found, while the status of the other inhabitants (around 9000) remained unclear.

Persistent rain, as well as rock slides and a layer of mud coating on the main roads, such as the one above, hindered rescue officials' efforts to enter the stricken region.

On 14 May, 1300 rescuers reached Yingxiu Town at the earthquake epicenter. Communication with the major town of Wenchuan was partly revived (in the afternoon), and 300 pioneer troops reached the main town of Wenchuan that night (about 23:30). Meantime, 15 Special Operations Troops, along with relief supplies and communications gear, parachuted into inaccessible Maoxian County, northeast of Wenchuan.

The Internet was extensively used for passing information to aid rescue and recovery in China. For example, the official news agency Xinhua set up an online rescue request center in order to find blind spots of disaster relief operations. After discovering that rescue helicopters had trouble landing in the epicentral area in Wenchuan, a student proposed a landing spot online, and it was used as the first place for the helicopters to touchdown.

Volunteers also set up several websites to store contact information for victims and evacuees.

5.3 The breadth and diversity of donors

The disaster relief received numerous monetary donations and other forms of aid from across the globe, which exceeded 456.9 million dollars.

Donations from countries and regions

Table 5.1 Donations from countries and regions (from http://en.wikipedia.org/wiki/Reactions_to_the_2008_Sichuan_earthquake)

	Amount donated (USD)	Material aid provided	Statements and condolences
Algeria	1,000,000		
Australia	4,500,000		Prime Minister Kevin Rudd: "On behalf of all Australian people, I express deep sympathy."
Bangladesh	5000	10 tons of relief materials	President expressed deep sympathy to the victims.
Belgium	1,026,349		

(continued)

	Amount Donated (USD)	Material Aid provided	Statements and Condolences
Bosnia and Herzegovina	70,000		
Botswana	150,000		
Brazil		Humanitarian aid in the form of food, blankets and tents	
Cambodia	10,000		
Canada	1,000,000	Medical team	"Canada will continue to help bring relief to those affected and provide support as people rebuild their lives and communities."
Denmark	750,000		
Estonia	49,000		
Finland	1,200,000	8000 tents, 1000 more army tents, 30,000 blankets	expressed its deepest condolences for the victims of earthquake and to the Chinese People
France	385,500	Tents, sleeping bags, blankets, tarpaulins, cooking kits and other materials	"France supports the Chinese people in this difficult moment."
Germany	31,000,000		"The German government is ready to provide speedy assistance."
Greece	310,000		"Greece shares the sorrow of the Chinese people and its mourning."
India	5,000,000		"India stands ready to offer any assistance that may be needed at this difficult time."
Indonesia		8 tons of medicines, 6 tons of food, 17 tons of platoon tents	"Indonesia prays for the people and the government of China as they overcome this disaster."
Ireland	1,550,000		
Italy	2,500,000		
Japan	9,600,000		"We hope that the Sichuan people can soon get on with their lives and rebuild their homes."
Kazakhstan		Food, tents, warm clothes and items of emergency to the amount of 3.6 million dollars	
Kyrgyzstan		120 tons of relief materials	
Laos	5,500,000		" We hope our support will help China in this most unfortunate time."
Lithuania	90,000		"Lithuania stands in solidarity with the people of China."

(continued)

		Amount Donated (USD)	Material Aid provided	Statements and Condolences
	Malaysia	1,500,000	4625 tents	
	Mauritius	300,000		
	Mongolia	50,000		
	Morocco	1,000,000		"We wish the injured a speedy recovery."
	New Zealand	500,000		
	D.P.R. Korea	100,000		
	Norway	3,900,000		
	Pakistan	50,000	500 blankets, 300 tents, 1000 plastic mats, 3 tons of bottled water and four tons of medicines	
	Philippines	450,000	Medical team	
	Poland	100,000		
	Portugal	150,000	Kitchenware, sanitary ware and food	
	Romania	860,000		
	Russia		30 tons of relief material from Russia arrived in Chengdu on 14 May 2008, becoming the first batch of international aid to reach China, with a 49-member rescue team	
	Samoa	100,000		
	Saudi Arabia	60,000,000		
	Senegal	500,000		
	Serbia	315,000	155 tents (for 2500 people)	"We sympathize with them at this difficult time."
	Singapore	200,000 (government) over 8,500,000 (private/local residents)	a 55-member rescue team; government sends $200,000 worth of medicines, drinking water, water purifying tablets, tents, groundsheets, blankets, sleeping bags and food	"We sincerely hope that the affected regions will return to normality in the shortest possible time."
	Slovakia	1,500,000	23 tons of rescue material	
	Slovenia	154,000		
	South Africa	200,000		
	R.O. Korea	5,000,000	a 41-person rescue team	"We wish for swift rescue and recovery operations in China."
	Spain	1,500,000		
	Switzerland	380,000		
	Tajikistan	100,000		

(continued)

	Amount Donated (USD)	Material Aid provided	Statements and Condolences
Thailand	920,131		
Tonga	50,000		
Turkey	2,000,000		
Turkmenistan		40 tons of relief materials	
Ukraine	1,007,000		
Venezuela	1,000,000		
United Kingdom	2,000,000		"I offer condolences and my profound sympathy to the Chinese people." —Queen Elizabeth
United States	4,877,598	Emergency relief supplies, and recovery equipment	"American people are with the Chinese people, the United States stands ready to help in any way possible."—George W. Bush
Vietnam	220,000		
Hong Kong China	38,400,000 (government) 128,000,000 (local residents)	A search and rescue team of 15 firemen, the 20-member team was equipped with approximately four tones of equipment, including life detectors and masonry cutting machines	
Macau China	15,300,000 (Government) 25,000,000 (local residents)		
Taiwan China	65,000,000	150 tons of supplies	

Donations from NGOs

On 13 May the Tzu Chi Foundation from Taiwan was the first force to join the rescue effort. A direct chartered cargo flight flew from Taiwan to Chengdu taking some 100 tons of relief supplies donated by the Tzu Chi Foundation and the Red Cross Society of Taiwan to the affected areas. A rescue team from Taiwan Red Cross took another direct chartered flight from Taipei to Chengdu in the afternoon of 16 May. The Amity Foundation had then already began relief work in the region and had earmarked 143,000 dollars for disaster relief.

Because of the magnitude of the disaster, and the media attention on China, foreign nations and organizations immediately responded by offering

Figure 5.15 A transport plane carrying 30 tonnes of relief material from Russia arrived in Sichuan's provincial capital Chengdu on 14 May 2008, becoming the first batch of international aid to reach China. Another 100 tonnes of goods arrived on three flights in the following days. In addition, a 49-member rescue team was sent to assist the rescue effort.

Figure 5.16 Poster of Wenchuan earthquake. "济" is a popular character in the Chinese language. The upper part of "济" is Wen of Wenchuan, and the lower part of "济" is chuan of Wenchuan. Wenchuan is made of these two separate characters in the Chinese language, and the ingenious combination of the characters, forms "济", which is a very positive word meaning "assistance", "help" and "happy" (design by Luo Rong, from http://together.chinavisual.com/).

condolences and assistance (Figures 5.15, 5.16). UNICEF reported that China formally requested the support of the international community to respond to the needs of affected families (Figure 5.17).

Francis Marcus of the International Federation of the Red Cross praised China's rescue effort as "swift and very efficient". But he added the scale of the disaster was such that "we can't expect that the government can do everything, and handle every aspect of their needs".

Figure 5.17 Local students run past a tent donated by the United Nations at a temporary school in Yinghua Township, in China's southwestern province of Sichuan on 11 November 2008. Six months after thousands of school children lost their lives in the Sichuan earthquake, psychological counselling remains a dire need for families here (photo by Liu Jin, AFP, Getty Images).

The breadth and diversity of donors

Following the earthquake, cash donations were made by people from all over mainland China, with collection booths set up in schools, at banks, and around gas stations. People also donated blood, resulting in long line-ups in most major Chinese cities. Many donated cash through text messaging on mobile phones to accounts set up by China Unicom and China Mobile. Online donations were then completely new and had not been used in China before.

The aid effort generated a remarkable breadth and diversity of donations, including the most extensive mobilization so far of compatriots of China in Taiwan, Hong Kong and Macao, where overseas Chinese gave generously (Figures 5.18, 5.19). Mediacorp, the main TV station of Singapore, staged a charity show on 25 May 2008, raising over 7.3 million dollars.

United Daily News reported that the top ten richest people in mainland China had together donated a little over 32.5 million yuan (4.6 million dollars) by 13 May, the second day after the earthquake. Many factories and industries, both private and state-owned, also announced donations of large amounts. Individual donors and enterprises included:

- Wang Laoji, a Chinese soft drink company, 100 million yuan
- Li Ka-shing, Chairman of Cheung Kong Holdings Limited, 130 million yuan
- Run Run Shaw, Chairman of Hong Kong Television Broadcasts Limited, 100 million HK dollars
- Wang Yung-ching, Founder of Formosa Plastics Group, 100 million yuan
- Nokia, 100 million yuan
- Samsung, 100 million yuan

Figure 5.18 20-year-old panda Qi Hao, a survivor of the May Sichuan earthquake gets a thorough physical examination in Fuzhou, southeastern China's Fujian Province on 30 Oct 2008. Qi Hao was transferred to the southern province four months ago after its home the Wolong Giant Panda Reserve Center in Sichuan was devastated by the earthquake (photo by AP).

Figure 5.19　Pupils paint sale for Wenchuan earthquake, students of primary school, Beijing.

Foreign governments, international organizations, and foreign friends all lent a helping hand, and up to 12 February 2009, the total donation from home and abroad was 41,742 million yuan (about 5.1 billion dollars).

5.4　Voluntary contributions

In November 1997, the United Nations General Assembly proclaimed 2001 as the International Year of Volunteers (IYV). United Nations Volunteers (UNV) was designated as the international coordinating organisation. With its main objectives of increased recognition, facilitation, networking and promotion of volunteering, the IYV provided a unique opportunity to highlight the achievements of millions of volunteers worldwide and it encouraged more people to engage in volunteer activity. As UN Secretary-General Kofi Annan said, "Volunteers can help transform all our societies, for the benefit of all people. But for this to happen, societies need to promote voluntary work as a valuable activity, and facilitate the work of volunteers at home and abroad."

Voluntary organizations in China are relatively new; volunteer work began with university students, and gradually extended to entrepreneurs and ordinary citizens. For example, the Youth Volunteer Corps of Sichuan University was officially founded in 1999; in 2004 it was reorganized into Sichuan University Young Volunteers Association. Using "based on campus, serve society" as its guiding ideology, the association members formed the vanguard of youth volunteers and have developed the spirit of volunteer service—"dedication, friendship, mutual aid and progress." More and more excellent young people have been attracted to take part in volunteer activities. Consequently, great achievements have been made in all areas of volunteer service on and off the Sichuan campus. Each year over nine thousand new volunteers register to join the association. At present, it has more than thirty thousand registered volunteers, 76 longstanding special volunteer service projects and 71 service bases, and these numbers continue to rise every year. There are currently 30 college-level youth volunteer groups, 18 community youth volunteer groups, 3 professional volunteer groups (Sunshine Mind volunteers, Library Volunteers, Museum Volunteers) and 5 assistance centers (Legal Aid Center, Medical Assistance Center, Cultural Assistance Center, Technological Aid Center and Volunteer Service Sports Center).

Sichuan University Young Volunteers Association is now the most participatory student organization for public service with the widest involvement on the campus, enjoying high social visibility. Volunteers serve the community needs as a positive response to the state's appeals. A lot of much-welcomed service work is carried out daily in areas like campus services, helping the poor, community building, unpaid blood donations, helping the elderly and the disabled, and in particular, emergency work in Wenchuan earthquake disaster relief and post- earthquake reconstruction.

An individual voluntary response

Chen Guangbiao is an entrepreneur, and also an enthusiastic volunteer.

Chen became aware of the earthquake during a board meeting in Wuhan, Central China. He cut the meeting short and headed straight for the stricken area, leading a team of 120 volunteers equipped with 60 pieces of heavy-duty construction machinery. 24 hours later, Chen and his team were on the ground, working in the affected area (Figure 5.20).

Figure 5.20 Chen Guangbiao—an enthusiastic volunteer, an entrepreneur (Chairman of Jiangsu Huangpu Co., Ltd.).
(a) Carrying earthquake victims;
(b) leading a team of 120 volunteers equipped with 60 pieces of heavy-duty construction machinery.

Their emergency response was almost as prompt as that of the Chinese military. They continued their work for 54 days, during which they buried about 6000 bodies and rescued 131 people trapped in debris. In addition, Chen donated 7.9 million yuan (about 1.13 million dollars) to rescue and relief operations, along with 23,000 radio sets, 3300 tents, 8000 sets of stationery, 1000 television sets and 1500 electric fans. Soon, the private industrialist came to be known as "Elder Brother Biao" to people in the afflicted area. Premier Wen Jiabao, who was in Sichuan directing the rescue and relief operations, met with Chen, and said: "I salute you for what you are doing."

Since founding his business in 1998, Chen has strongly supported charitable works with donations, in cash and in kind valued at more than 475 million yuan (about 69 million dollars). In 2008, he was named "China's No. 1 philanthropist" by the Ministry of Civil Affairs. He often compares wealth to water: "When you just have one cup of water," he has said, "you may keep it for yourself. But when you have more than enough for yourself you should share it with others: for example, when you own a river."

Rewards of voluntary work

The volunteering spirit has spread amongst the Chinese people in recent years, especially among young people (Figure 5.21). Volunteering is of tremendous benefit to society and those in need, and the Wenchuan earthquake was a good example. Hundreds of thousands of volunteers played an active role in Wenchuan quake-hit regions. They contributed daily necessities, and offered medical help and psychological help to the disaster relief work.

It must be noted too that volunteering was beneficial to the volunteers themselves: in volunteering, volunteers are exposed to a new environment, and they can learn how to work well in a team, and improve their interpersonal skills and organizational skills, all of which are critical for their professional education and development. Everyone is encouraged to take the opportunity to join voluntary work. It is not only good for society, but is also a chance for us to mature and learn.

Chapter 5 · Emergency response and rescue | 217

Figure 5.21 A little earthquake survivor and the patients who were injured in the Sichuan earthquake perform a sign language song at the rehabilitation centre of Sichuan Provincial People's Hospital on 7 May 2009 in Chengdu. Around 226 people wounded in the earthquake have received medical care at the Sichuan Provincial People's Hospital's rehabilitation centre, where 59 of them are still undergoing treatment (Photo by Li Feng, Getty Images).

Medical volunteers

On 15 May, the fourth day after the earthquake, Taiwan Buddhist Tzu Chi Foundation sent the first batch of relief medical supplies directly to Chengdu. The Foundation in Shanghai, Suzhou, Kunming, Fujian, Guangdong, Beijing and other places then supplied 16 volunteers who formed the first batch of disaster relief team members, flying to Chongqing to group together before driving to the disaster area. Many countries sent medical volunteers to Wenchuan earthquake areas (Figure 5.22).

A Disaster Relief Team is usually divided into four groups: a medical care group, a social care group, a cooking group, and an engineering group, for implementing multi-function, multi-angle professional disaster relief.

One medical group doctor observed that children could not attend school because of the earthquake, had nothing to do, and had no psychological care. Tzu Chi volunteers collected a group of small children to be volunteer translators of the Sichuan dialect, to assist clean-up operations, or accompany home visits. These children also provided information to the

Figure 5.22 Medical volunteers from Germany (photo by Lin Yiguang, Xinhua News Agency).

Disaster Relief team on the mother's role, the injuries requiring treatment, and who should take care of the elderly. One mother said that after the earthquake her son became very scared, very quiet, and did not want to eat. Once volunteers joined up with the children, the voices of children were heard once again, full of vigor and vitality.

During the emergency period, Tzu Chi established four service stations in different locations, providing hot food, which countered the danger and inconvenience of survivors cooking in tents.

Importance of psychological assistance

Besides saving lives, and healing physical injuries, psychological assistance to the victims become another focus of the relief work (Figure 5.23). Various societies and research institutions sent experts on missions to the region to carry out psychological intervention. Such large scale organized counseling to the victims, the first since the founding of the new Chinese state, is a concrete manifestation of social progress. After the earthquake, many young

Figure 5.23 Be strong to survive! 20 May: this surviving couple started a new life in the ruins of Beichuan County (photo by Zhao Qing, Shenzhen Evening News).

people were willing to help the people in Sichuan.

The heartfelt belief of every volunteer is that helping the people in Sichuan was very meaningful and worthwhile. The volunteers brought love and fellowship, and with that they gave hope to the people in Sichuan. They not only helped the local people rebuild their homes but also strengthened their hearts (Figure 5.24). The young volunteers were strengthened too, by their experiences. Through their experiences and the help that they gave to the survivors in Sichuan, they will go on to demonstrate to people outside the disaster area that life is fragile. And then these people will know how to cherish their own life.

Nowadays, more and more people kill themselves because they feel unhappy, and their relatives are deeply hurt when they die. In the earthquake, many people died and many people lost their relatives. The survivors suffered from the pain of their relatives and friends leaving them. Our life gives us a responsibility for others—this duty we must recognise. Being a volunteer teaches us the vital lesson of cherishing what we have.

The development of voluntary work is part of the growth of a mature civil society. From across the country more than million volunteers arrived in Sichuan (where 40% of students at university volunteered). Volunteer action was their own implementation of the maxim "dedication, friendship, mutual aid, and progress" common to voluntary work and society.

The earthquake caused a huge disaster, but it also consolidated and improved rescue systems, emergency response mechanisms, and fund-raising mechanisms, as well as mobilizing society. The experience gained is a valuable resource for the future of China's relief and social development efforts, and it has provided many useful insights.

Figure 5.24 Chengdu Sports Aid (CSA) is group of volunteers, Chinese and foreign, based in Chengdu, seeking to relieve some of the hardships of the Sichuan earthquake victims, through community sports days and other fun events (from the website of Sichuan Quake Relief which is a non-profit, humanitarian organisation dedicated to improving the lives of those affected by the 12 May 2008 Sichuan earthquake, http://sichuan-quake-relief.org/projects/chengdu-sports-aid/).

The rescue efforts performed by the Chinese government were praised by critical western media, especially in comparison with Myanmar's blockage of foreign aid after Cyclone Nargis, as well as China's previous performance at the time of the 1976 Tangshan earthquake.

References

Chen Y, Chen Q F, Liu J, et al. 2002. Seismic Hazard and Risk Analysis: A Simplified Approach. Beijing: Science Press.
China Earthquake Administration. 2008. 2008 8.0 Magnitude Earthquake Rescue Documentary. Beijing: Seismological Press. (in Chinese)

Figure 6.1　New Beichuan after two years reconstruction work (photo by Guo Xun).

Reconstruction of Wenchuan

223 / Outline of reconstruction
229 / Counterpart assistance
235 / Solid schools, sweet hopes
241 / New Wenchuan after the earthquake

6.1 Outline of reconstruction

The major earthquake which struck Wenchuan at 14:28 on 12 May 2008, took tens of thousands of lives, deprived millions of households of their homeland on which they had relied for generations, and destroyed in a instant the fortune accumulated by painstaking work over decades.

Post-Wenchuan earthquake restoration and reconstruction is an arduous task. Confronted with the difficult situation of such a wide disaster-affected area, such a large disaster-affected population, a difficult natural environment and severely damaged infrastructures, post-quake restoration and reconstruction is extremely strenuous and challenging. It is the most extensive mobilization, and the largest reconstruction project since the foundation of new China.

The project of restoration and reconstruction will last 3 years, and the total capital demand for restoration and reconstruction is calculated approximately as 1000 billion yuan (135 billion dollars).

In fact, the reconstruction worked very quickly, two years after earthquake, Wenchuan earthquake reconstruction has made great progress (Figures 6.1, 6.3, 6.4)

Overall planning

In order to perform a good restoration and reconstruction job in an authoritative, orderly and efficient way, the State Council created a reconstruction plan.

The Planning Group for Post-Wenchuan Earthquake Restoration and Reconstruction under the State Council was established soon after the earthquake, under the joint leadership of the National Development and Reform Committee (NDRC), and the People's Government of Sichuan Province, Ministry of Housing and Urban-Rural Development (MOHURD).

A draft of "The State Overall Plan for Post-Wenchuan Earthquake Restoration and Reconstruction" was issued to the public for their opinion on 18 August 2008, and the final official edition was issued on 28 August 2008 (for extracts see Appendix 4).

This paragraph appeared at the beginning of the overall plan:

This plan is formulated in order to extend deepest condolences for the victims of the Wenchuan earthquake, reinforce our sympathy to all disaster-affected people, and express the sincerest gratitude to those who have been concerned about and given support to the earthquake fighting and disaster relief efforts and restoration and reconstruction of the affected areas.

The plan then gave the brief summary of the disaster which is given at the beginning of Chapter 1. The disaster losses quoted in the plan are particularly severe. Tables 6.1 & 6.2 shows the disaster-induced losses in the zones covered by the plan.

The Wenchuan earthquake affected 417 counties (cities, districts) of 10 provinces (autonomous regions and municipalities) such as Sichuan, Gansu, Shaanxi, Chongqing, Yunnan Provinces, etc., covering approximately a total area of 500,000 km^2. The scope of the overall plan includes 51 counties (cities and districts) in the hard-hit and extremely hard-hit disaster areas of Sichuan, Gansu and Shaanxi Provinces, covering a total area of 132,596 km^2, involving 14,565 administrative villages in 1271 towns and townships, with a total population of 19.867 million by the end of 2007, and a gross regional product of 241.8 billion yuan (Figure 6.2).

Table 6.1 Disaster-induced losses in the areas of the reconstruction plan

Item	51 Counties (cities, districts)
Direct economic losses (hundred million yuan)	8,437.7
Damaged highway (km)	34,125
Damaged reservoirs	1,263
Damaged transmission lines (km)	61,524
Damaged substations above 35 kV	250
Damaged schools	7,444
Damaged medical institutions	11,028
Collapsed rural residences (10,000 m^2)	10,709
Severely damaged rural residences (10,000 m^2)	9,432
Collapsed urban residences (10,000 m^2)	1,877
Severely damaged urban residences (10,000 m^2)	5,836

Data source: Appendix 4, modified.

Table 6.2 Planned scope

Province	County (city and district)	No.
Sichuan	Wenchuan, Beichuan, Mianzhu City, Shifang City, Qingchuan, Maoxian, Anxian, Dujiangyan City, Pingwu, Pengzhou City, Lixian, Jiangyou, Lizhou (District of Guangyuan City), Chaotian (District of Guangyuan City), Wangcang, Zitong, Youxian (District of Mianyang City), Jingyang (District of Deyang City), Xiaojin, Fucheng (District of Mianyang City), Luojiang, Heishui, Chongzhou City, Jiange, Santai, Langzhong City, Yanting, Songpan, Cangxi, Lushan, Zhongjiang, Yuanba (District of Guangyuan City), Dayi, Baoxing, Nanjiang, Guanghan City, Hanyuan, Shimian, Jiuzhaigou	39
Gansu	Wenxian, Wudu (District of Longnan City), Kangxian, Chengxian, Huixian, Xihe, Liangdang, Zhouqu	8
Shaanxi	Ningqiang, Lueyang, Mianxian, Chencang (District of Baoji City)	4

Figure 6.3 Photo taken on 12 May 2010 shows a view of the ruins of old Hanwang Town in Southwest China's Sichuan Province. A hand-over ceremony of 22 reconstruction projects in Hanwang took place on 12 May 2010, the second anniversary of the Wenchuan earthquake. The projects were sponsored by the city of Wuxi in East China's Jiangsu Province. A new Hanwang is growing up beside the ruins, thanks to rebuilding works in the past two years (courtesy of Zhang Hongwei).

Figure 6.4 A new stadium has been built at Wenchuan County, two years later after large earthquake (photo by Dong Yong, Sichuan Earthquake Administration).

Central finance shall provide funds for post-quake restoration and reconstruction, to a level of 30% of the total. In 2009, the Central Government's public investment was 924.3 billion yuan (135.33 billion dollars). "Of this, 14 percent was invested in post-Wenchuan earthquake recovery and reconstruction," said Chinese Premier Wen Jiabao in his government work report to the annual session of the National People's Congress, the country's top legislature, in March 2010.

The funds for restoration and reconstruction shall be collected through various channels: local government allocation, counterpart assistance, social donations, domestic bank loans, capital market financing, foreign emergency loans on favorable terms, urban and rural self-possessed and self-collected capital, self-possessed and self-collected capital of enterprises, innovation financing, etc.

On 5 December 2008, the World Bank provided a loan of 710 million dollars to China for the Wenchuan Earthquake Recovery Project (WERP). The project will finance post-earthquake reconstruction and recovery in infrastructure, health and education sectors in Sichuan and Gansu Provinces and will assist both provinces in laying the foundation for the longer-term sustainable economic recovery of areas severely affected by the Wenchuan earthquake.

The United Nations also played a modest but substantive role in "Building Back Better". In response to the earthquake, the UN System in China mobilized more than 70 million dollars for emergency relief, early recovery and reconstruction to assist people and affected communities in Sichuan, Gansu and Shaanxi Provinces.

6.2 Counterpart assistance

Less than a month after the earthquake, the State Council promulgated the Post-Wenchuan Earthquake Restoration and Reconstruction Counterpart Assistance Plan. The plan requested that China's 19 eastern and central provinces and municipalities pair up with the hardest-hit regions. This model for aid has been executed smoothly in Sichuan (Table 6.4).

New mechanism of post-earthquake reconstruction

Counterpart assistance is a resource allocation and regional cooperation mechanism with significant Chinese characteristics. Under this mechanism, developed regions use their fiscal revenue to provide financial support to recipient regions.

This mechanism was not invented in response to the Wenchuan earthquake. In the 1970s and 1980s, there were policies in China that designated economically developed eastern regions such as Shanghai and Jiangsu Province to provide counterpart assistance to Tibet, Xinjiang, the Three Gorges Reservoir area and other poor or extremely poor regions. After the Wenchuan earthquake, various provinces actively gave a helping hand and provided all kinds of assistance to the disaster-stricken regions. The Central Government quickly put this voluntary assistance in order, and incorporated the disparate elements into its national plan for post-disaster reconstruction. On 11 June 2008, the State Council decided: 19 provinces (cities) shall offer assistance of not less than 1% of their last ordinary budget revenues to their 24 counterpart counties (cities, districts) in Sichuan, Gansu and Shaanxi Provinces.

The 19 provinces, municipalities and autonomous regions, which were committed to providing help for quake-hit areas, also invested more than 77 billion yuan (11.27 billion dollars) in the post-quake reconstruction and 3 356 projects were confirmed.

Relief for previous serious disasters primarily took the form of direct fiscal transfers from the Central Government. Introducing the counterpart assistance mechanism into post-disaster restoration and reconstruction is a useful new tool for disaster relief with novel Chinese characteristics.

Figures 6.5, 6.6, 6.7 & 6.8 show the Wenchuan County at earthquake and after reconstruction respectively, this is result of counterpart assisstance of Guangdong Province to Wenchuan County.

Compared with fiscal transfer from the Central Government, counterpart assistance not only provides funds to the earthquake-stricken area, but also provides strong material support, advanced management ideas and highly-qualified talents. The areas in Sichuan that were hit worst in the earthquake were relatively closed off and economically

Table 6.4 Counterpart assistance list

1. Shandong Province	–	Beichuan County, Sichuan
2. Guangdong Province	–	Wenchuan County, Sichuan, and hard-hit areas in Gansu Province
3. Zhejiang Province	–	Qingchuan County, Sichuan
4. Jiangsu Province	–	Mianzhu City, Sichuan
5. Beijing	–	Shifeng City of Sichuan Province
6. Shanghai	–	Dujiangyan City of Sichuan Province
7. Hebei Province	–	Pingwu County, Sichuan
8. Liaoning Province	–	Anxian of Sichuan
9. Henan Province	–	Jiangyou City, Sichuan
10. Fujian Province	–	Pengzhou City, Sichuan
11. Shanxi Province	–	Maoxian County, Sichuan
12. Hunan Province	–	Li County, Sichuan
13. Jilin Province	–	Heishui County, Sichuan
14. Anhui Province	–	Songpan County, Sichuan
15. Jiangxi Province	–	Xiaojin County, Sichuan
16. Hubei Province	–	Hanyuan County, Sichuan
17. Chongqing	–	Chongzhou City, Sichuan
18. Heilongjiang Province	–	Jiange County, Sichuan
19. Tianjin	–	Hard-hit areas in Shaanxi Province

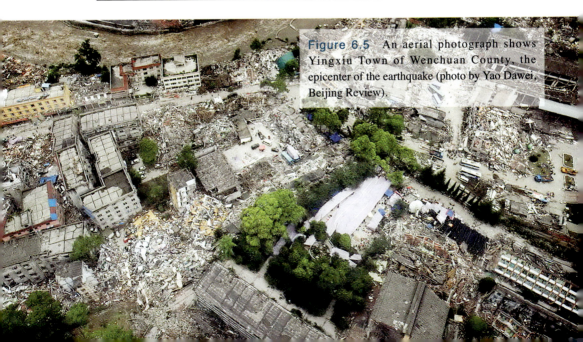

Figure 6.5 An aerial photograph shows Yingxiu Town of Wenchuan County, the epicenter of the earthquake (photo by Yao Dawei, Beijing Review).

Figure 6.6 More than 200 families of Zhangjiaping Village, Yingxiu Town, will move into new houses that have been built by Guangdong Province's Dongguan City, an assistance program partner (photo by Huang Runqiu).

underdeveloped. The objective behind restoration and reconstruction is not simply to return these areas to their previous level. The Central Government requires that reconstruction should not simply mean giving money to disaster-stricken areas or building houses or roads. Rather, it should gear aid toward constructing an all-around and relatively well-off society, promoting economic development, and speeding up the rebuilding of the areas into a beautiful homeland.

For more than a year, all provinces and municipalities providing assistance have been highly innovative and creative in the reconstruction of their counterpart disaster-stricken areas. They have brought in new ideas and new technologies, guaranteed both funding payments and the timely completion of projects, and provided local training in technology and management. In many areas, the assisting workforce has positively

Figure 6.7 A new artificial lake built at Wenchuan County, two years after earthquake (photo by Dong Yong, Sichuan Earthquake Administration).

influenced the mindset of local residents.

Government-administered post-disaster reconstruction can mobilize broad social support, invest very large amounts of resources in a short time, swiftly restore production and daily life in disaster-stricken areas and enable victims in disaster-stricken areas to rise up from the ruins as quickly as possible. Compared with other initiatives, the government-administered mechanism is more efficient, powerful and effective.

In a large developing country like China, there are major and obvious regional differences. In terms of the government's long-term strategy, the counterpart assistance mechanism in post-Wenchuan earthquake reconstruction optimizes allocation of resources, conforms to its principle of coordinated regional development, and sets up a model for future relief work in serious future disasters. It helps people in disaster areas solve many urgent problems in their lives. As of the end of 2009, 19 assistance-providing

Figure 6.8 Frozen In Time: The photo shows a Wenchuan County memorial park, where the centerpiece is a tower surmounted by a clock which stopped at the moment of the 2008 earthquake.

provinces and municipalities had set up 3105 counterpart assistance projects in Sichuan. Of these projects, 91% were under construction and 50% had already been completed.

Wenchuan was paired with Guangdong Province, which has one of the fastest-growing economies in China. The plan drawn up by Guangdong states that it will spend 8.2 billion yuan (1.2 billion dollars) on reconstruction, mainly in building rural and urban housing, disaster shelters, roads and medical, urban water supply, social welfare, leisure and sports and agricultural facilities. Work crews completed all projects on 3 December 2009, and most of the Wenchuan population that was affected by the disaster have moved into permanent housing.

From counterpart assistance to counterpart cooperation

The counterpart assistance mechanism may create new opportunities for economic development in western China. Counterpart assistance funds play an important role in helping restoration, reconstruction and sustained development in the quake-hit zones. They supply much needed funding for post-disaster reconstruction, and help promote social reform by increasing regional cultural exchanges and communications and bringing in new ideas for development of the quake zones. They also improve industrial transfer and structural adjustment, industrial development, and regional cooperation and exchange. They bring in advanced ideas and technologies, and open new markets and new channels for cooperation and communication.

The counterpart assistance plan is a great initiative pioneered by China. It is a unique Chinese model that combines socialist theory with Chinese characteristics with longstanding practices in post-quake restoration and reconstruction. Professor Hitoshi Taniguchi from the Nagoya Institute of Technology said, "counterpart assistance could not be achieved by any other country but China".

Counterpart assistance is mainly a government action, which would be hard to replace by free market functions. After the first stage of counterpart assistance, the next stage should be cooperation between the counterpart provinces/municipalities and the quake zones. This transition has already taken place in some parts of Sichuan Province. The whole project is being reoriented from relief assistance to development assistance, and shifted from counterpart assistance to counterpart cooperation (Liu, 2010).

6.3 Solid schools, sweet hopes

Schools in reconstruction

According to the statistics (Table 6.1), 7444 schools were damaged during the Wenchuan earthquake.

The possible reasons for a disproportionate number of school collapses are:

• Many school buildings were old and altered after construction to accommodate increased pupil numbers;

• Building control by government departments may have been of poor standard;

• Some buildings had large spans which lacked structural support;

• Precast concrete building systems were used, with poor continuity at joints.

What can be learned from the damage to schools?

1. Schools are the most important public facility. Schools are places for children. Their collapse not only causes many casualties, but can lead to social instability because so many families are involved. Also, many people have recommended designing public facilities like schools and hospitals as multi-functional buildings which can serve as shelters or rescue and relief headquarters in case of emergency. Therefore, government must also make school building codes more stringent and raise seismic resistance standards for schools. The provision in the overall post-Wenchuan earthquake plan is: the seismic fortification intensity for schools should be one degree higher than the basic intensity.

2. In addition to improving building codes, government should guarantee they are implemented, in order to ensure the quality of buildings, particularly public buildings designed for large numbers of people. California is a good example as it relies on an independent agency to supervise school construction, from reviewing the construction plans at the outset to continuously inspecting the construction sites. Public accountability and expenditure auditing can be effective tools for quality supervision.

3. The number of damaged buildings in the area affected by the Wenchuan earthquake is too great to envisage rebuilding them all in a short time; a feasible, cost-effective solution is to identify the most vulnerable buildings for retrofitting/renovation and then expand the scope of this project until all buildings throughout the country are retrofitted. The conventional reactive seismic strategy should be changed to a proactive one. The focus of disaster management should be shifted from recovery-oriented to mitigation-oriented.

It is inspiring to see that the Wenchuan No. 2 Primary School has adopted mature and effective seismic technology such as base isolation, together with quality materials, to guarantee the seismic safety of public facilities such as schools. This School uses modern architecture in the Qiang ethnic minority style. The vibration-isolation cushion in the building can be seen through windows to make the local people aware of the technology. Cushions can be used for up to 50 years before being replaced. The technology was developed in a Guangzhou University laboratory.

Sichuan Provincial Education Department said 2997 schools in 39 of the worst hit counties were completed or nearing completion, accounting for 99.83% of the 3002 schools that needed to be rebuilt. Of those structures 388 are still under construction while 2609 had been completed by 5 May 2010 (Figures 6.9, 6.10).

Figure 6.9 The athletic field and gymnasium of New Beichuan middle school. 200 million yuan was invested in its construction; the construction area is 72,000 m^2 (photo by Dong Yong, Sichuan Earthquake Administration).

Figure 6.10 Buildings at the Wenchuan No. 2 Primary School compound in Sichuan Province have been rebuilt with new vibration-minimizing cushion technology (Photo by Dong Yong, Sichuan Earthquake Administration).

Some schools in the Wenchuan area are now "twinned" with schools in Beijing. In future, Beijing will send teachers to the school year teaching program, and Wenchuan schools will send teachers to Beijing for teacher training in exchange.

Lessons learned from the Wenchuan earthquake triggered a substantial revision of the Earthquake Act (see Appendix 3). One addition to the Act is a requirement that the design and construction of densely occupied buildings such as schools and hospitals meet higher standards than the seismic fortification specifications for regular buildings in the same area. There is hope that seismic risk mitigation will thus continue to improve across the whole country.

Hospitals and other public facilities

Earthquake caused serious damage to public service facilities; some were almost "paralyzed". Education, health, employment and other public services became the most urgent post-disaster concerns in public aspirations for a better livelihood. In response, the State Plan put earthquake reconstruction for protection of the people's livelihood as the basic starting point, giving priority to the renewal of schools, hospitals and other public services and facilities (Figure 6.11).

Figure 6.11 New Wenchuan people's hospital (photo by Dong Yong, Sichuan Earthquake Administration).

A total of 269 million yuan (39.6 million dollars) was designated to rebuild medical facilities. The total area of newly built facilities is 60,000 m^2, which is 1.6 times that which was available before the earthquake. There are 462 new in-patient hospital beds, which is an increase of 40% compared with the number before the earthquake. Guangdong has also sent 1544 medical staff members to Wenchuan to provide medical services, and train doctors, nurses, medical equipment operators and hospital management personnel.

After nearly two years of post-disaster reconstruction in the disaster areas, public services have not only returned to normal, but the hardware and software levels before the earthquake have had a qualitative upgrade. Many new and widespread employment assistance centers have brought skills and jobs to the earthquake survivors, bringing "sweet hopes" (Figure 6.12).

Brief details of the reconstruction of facilities including water, transportation, social welfare, leisure and sport, rural public service, logistics facilities for agricultural products, disaster shelters, etc., are as follows:

Water facilities: Before the earthquake, Wenchuan had only two water treatment plants, with a combined supply capacity of 7000 tons per day. Guangdong spent 221 million yuan (32.5 million dollars) on 47 projects to repair and upgrade the urban water supply infrastructure. Daily capacity has risen to 18,100 tons.

Transportation facilities: The earthquake destroyed 340.3 km of rural roads and 80 km of urban roads. About 60% of the county's concrete and asphalt roads and three highway bridges were destroyed. Guangdong donated 1.53 billion yuan (224 million dollars) to restore roadways, resulting in the rebuilding of 208 km of urban roads and 252 km of rural roads (Figure 6.13).

Social welfare facilities: Wenchuan had only one social assistance and welfare service center before the earthquake, with a total area of 3600 m^2 and 150 beds. Guangdong allocated 72 million yuan (10.6 million dollars) for social welfare projects. Two social welfare centers have been built with a total area of 13,849 m^2 and 540 beds.

Leisure and sports facilities: Before the earthquake, Wenchuan's libraries, leisure centers and bookstores were small in number and size and the county had no public sports venue. Guangdong spent 172 million yuan (25.3 million dollars) on 28 leisure and sports facilities projects.

Rural public service facilities: Guangdong donated 71.3 million yuan (10.5 million dollars) to build rural public service centers, with a total area of 46,400 m^2.

Logistics facilities for agricultural products: Guangdong allocated 67.3 million yuan (9.9 million dollars) to build 10 agricultural products markets in 13 Wenchuan townships. Ten markets have been completed after an investment of 49.8 million yuan (7.3 million dollars).

Disaster shelters: The earthquake's environmental destruction in Wenchuan has made the county prone to secondary disasters, such as landslides, mudslides and the collapse of river embankments. Guangdong earmarked 542 million yuan (79.7 million dollars) to build new disaster shelters. Nineteen shelter facilities have been completed at a cost of 124 million yuan (18.2 million dollars).

Figure 6.12 A high-speed train first ran from Chengdu to Dujiangyan in Southwest China's Sichuan Province on 10 May 2010. The railway began operating on 12 May as part of efforts to rebuild Sichuan's transportation network.

Figure 6.13 New road at New Wenchuan (Photo by Dong Yong, Sichuan Earthquake Administration).

6.4 New Wenchuan after the earthquake

Two years of hard and painstaking work have passed. Today, how have circumstances been transformed for those whose lives were devastated by the earthquake? Figure 6.14 showed the time table for reconstruction.

All be housed

According to Sichuan Province statistics, the earthquake damaged 3.7 million rural houses, of which 1.5 million needed replacement, and 2.2 million required repair or reinforcement. For urban housing units, 259,000 needed replacement, and 1.3 million required repair or reinforcement.

After the earthquake, all levels of government quickly issued a post-disaster urban and rural housing reconstruction program, to provide subsidies to affected families. The financial assistance given was 25,000 yuan per household to urban families, and 20,000 yuan per household in rural areas. Farmers in self-built housing and requiring resettlement received 2000 yuan per household for construction of transitional housing. On 30 April 2010, the reconstruction of rural houses and urban houses was complete. Maintenance

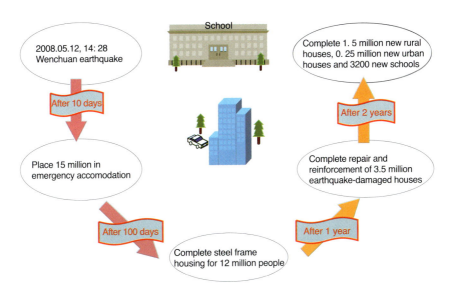

Figure 6.14   Time table for reconstruction.

and reinforcement of 1,348,100 urban houses had also been completed (Figures 6.15, 6.16).

After the earthquake, Chengdu introduced a post-disaster revision of its urban and rural housing policy, allowing the introduction of social funds for disaster victims to rebuild rural housing. For example, if a farmer has building sites, and a city person has money, they can form a construction partnership. The city person is responsible for financing and constructing two buildings on these sites, so that each partner has one. In this way, farmers do not have to provide cash to get a new building.

Life goes On

The huge losses constitute a big challenge to post-quake reconstruction efforts. The Central Government pledged to provide quake victims with

Figure 6.15 New housing in new Beichuan County, constructed by Shandong Province: 1.22 million m^2 11,084 homes.

a stable life within three months. For this purpose, 70 billion yuan (about 10 billion dollars) was allocated from the state treasury in relief funds, and 19 economically developed provinces have been designated to help with infrastructure rebuilding in the severely hit counties. Using personnel and technical assistance from all over the country, water and power supplies, and roads and telecommunication networks have been restored to pre-earthquake levels in most affected areas, for the time being.

While help was pouring in from many sources, the survivors did not stand idly by. Only days after the devastating quake, many Sichuan farmers returned to work, risking frequent aftershocks. They made great contributions to another bumper harvest for the province, which, according to the provincial department of agriculture, saw a year-on-year increase of 50 million kg in grain output and an increase of 150 million kg in oil seed output.

Figure 6.16 Photo taken on 12 May 2010 New Luobu Village of Wenchuan County on the second anniversary of the Wenchuan earthquake. The projects were sponsored by Guangdong Province (Photo by Dong Yong, Sichuan Earthquake Administration).

Official statistics show that, on 8 July 2008, 83% of large industrial enterprises in Sichuan and more than 90% of local commercial businesses had resumed operation after quake-caused halts. For example, Dongfang Turbine Co. Ltd., China's second biggest power-equipment maker, which is located in hardest-hit Deyang City, made its first post-quake generator delivery on 20 May.

What is more significant is that life is gradually returning to normal in the quake zone. More than three-quarters of homeless survivors have moved into temporary homes from tents. Schools have reopened. Tourist sites are again accessible. Restaurants are welcoming a growing number of diners (Yao, 2008).

Liu Qibao, Secretary of the Sichuan Provincial Committee of the Communist Party of China (CPC), at a review meeting of post-quake reconstruction on 8 May 2010 said that by 30 April, construction had started on 28,886 projects, accounting for 97.2% of the 29,704 projects slated for completion under the general reconstruction plan (Figure 6.17). Some 678.75 billion yuan (99.38 billion dollars) had been invested in rebuilding the quake-hit areas, accounting for 72.3% of the total planned investment.

Figure 6.17 New houses of Qiang ethnic minority groups, Jina Village of Wenchuan County.

"With the gradual completion of more reconstruction projects, especially the livelihood projects, the emphasis of post-quake reconstruction will shift to improving the development capacity of the quake-hit areas", Liu said.

All the people in Wenchuan region believe that after a three-year reconstruction period, people in affected areas will have far better living and business conditions than those before the earthquake struck (Figure 6.18).

The main lessons of the earthquake

Assessment of the effects of the Wenchuan earthquake has led us to the following main conclusions:

In seismically active areas, particularly those with a history of large earthquakes (which may have happened many tens, or even hundreds of years ago), seismic hazards have to be very carefully evaluated, so that so that the appropriate building codes for seismic resistance are applied.

Building standards and the seismic resistance of most buildings were shown to be severely deficient during the exceptionally strong shaking produced by the Wenchuan earthquake. However, where good building standards had been applied, such buildings stood up well to the earthquake. The seismic resistance of all buildings, particularly public buildings designed for large numbers of people, must be greatly improved in seismically active areas all over China.

Emergency measures approved by the government and implemented in Wenchuan worked well, involving teams from China and all over the world for rescue and relief, including volunteers, and new schemes for reconstruction, with twinning of affluent and disaster-stricken areas.

In the years following the Tangshan disaster, China has taken a more objective view of the possibility of earthquake prediction. Most Chinese seismologists believe that only a very few earthquakes which show a recognizable foreshock sequence have a prediction capability and a solution to the problem is far away. Meantime, they are not neglecting the possibility of recognizing foreshocks.

Earthquake-induced landslides caused many deaths and severe disruption of transportation systems, particularly roads and railways on steep hillsides. Well-engineered bridges and tunnels suffered much less damage, so their use should be adopted as far as possible in mountainous terrain to minimize the risk of damage due to landslides and subsidence.

Figure 6.18 Dancing to celebrate their new houses at Cigou Village of Wenchuan County (photo by Dong Yong, Sichuan Earthquake Administration).

References

Liu S Q. 2010. New opportunities for economic development. http://www.bjreview.com.cn/Post_Wenchuan/2010-02/11/content_246680.htm. Accessed 25 Apr 2011.

Yao B. 2008. Life goes on. http://www.bjreview.com/special/sichuan_earthquake/txt/2008-07/25/content_136498.htm. Accessed 25 Apr 2011.

Appendices

Appendix 1

Significant events of Wenchuan earthquake

5.12

- Earthquake at 8.0 hits South-West China
- Premier Wen Jiabao arrived in Sichuan earthquake-hit region
- China sends National rescue team to Chengdu
- Cell phone service affected in quake areas
- China Eastern Airlines cancels flights to quake area
- Water pipe broke in earthquake disaster areas
- Up to 8500 die in killer earthquake (Xinhua news)
- Rain to linger in quake-hit area of SW China

5.13

- USGS: China quake caused by motion on reverse fault
- 1300 rescuers reach epicenter
- Earthquake disrupts railway transportation
- Quake closes major highways in SW China
- Chengdu Shuangliu international airport re-opens
- Telecom services hit
- 14 Makeshift medical center in quake-hit area
- Strong aftershock rocks Chengdu
- Witnesses: Villagers debris, whole town leveled
- Health ministry sends 10 medical teams to quake-hit areas
- listed firms stop trading after quake
- Isolated quake epicenter appeals for emergency aid

- Rescue teams thwarted from entering epicenter
- At least 1000 students buried in China county worst hit by quake

5.14

- Major oil pipeline in China's quake region back in service
- Electric plant one of the worst hit
- Air, railways reopen, but roads still blocked
- Relief teams rescue 84 survivors in Sichuan
- Soldiers enter earthquake epicenter, save thousands
- Emergency measures for power supply
- 21.6 million bottles of mineral water shipped to quake-hit areas
- Disaster zones in urgent need of AB blood
- EU says ready to provide aid to earthquake-hit China
- Local governments donate 90 million yuan
- China allocates another 250 m yuan to quake relief fund
- 100 soldiers parachuted to cut-off quake area
- Army sends 70 medical teams to quake areas
- Soldiers hike to quake-buried Chinese villages
- PM orders reinforcements as toll rises to 14,866
- China stresses prison stability after quake
- 178 students confirmed dead in one school in Sichuan

5.15

- Dam near quake epicenter "structurally safe"
- Impacted dams getting repaired
- More than 50 tourists die in earthquake
- Hard-hit Hanwang picks up the pieces
- Wenchuan roads are still blocked
- Phones coming back to service at quake epicenter
- Injured airlifted to Chengdu from epicenter
- Trapped teenagers sang songs while awaiting rescue
- More Russian aids arrive in quake-hit Chengdu
- Evangelist sends best wishes, donation
- Japan to send relief team to China
- Beijingers weather downpour to donate blood

- 130,000 troops in operations throughout quake area
- Landslide hinders rescue efforts
- At least 270 students dead in quake in one school
- China's quake region imposes price controls

5.16

- Insurers not jolted by quake
- More than 20,000 survivors rescued
- 30 miners confirmed dead, more missing
- Beichuan County officials rescued 75 hours after quake
- President Hu inspects quake battered province
- Information from quake epicenter remains patchy
- Communications being restored in SW China
- Singapore sends relief team to China
- Adoption applications flood in
- Japanese rescue team starts working in Qingchuan
- Warm weather for quake zone; epidemic prevention urged

5.17

- German saved out of rubble 114 hours after quake
- Telcoms being restored, but aftershocks interfere
- Water purifiers sent to shelters
- Donations to China quake zone hit 6.023 billion yuan
- Bad weather hinders rescue of 14 Taiwan tourists
- Telecom disruptions hamper rescue effort

5.18

- China to investigate cultural relic damages in quake zones
- Three giant pandas missing from Wolong after quake
- Eleven quake-stranded Taiwan tourists rescued
- 7 rescued after 120 hours in debris
- 63 people rescued after five days in debris
- China dispatches 113,080 soldiers in quake rescue
- Rain hits part of China's quake areas
- 21 landslide dams threaten quake-hit areas

5.19

- Chinese diplomatic missions mourn quake victims
- PetroChina resumes oil supply to quake-hit areas
- Three bilingual websites set for quake information
- Woman rescued alive 150 hours after quake
- Companies donate over 3.5 billion yuan for quake relief
- HK to send assessment team to Sichuan
- 21 lakes formed, "no danger yet"
- Weather blocked immediate assistance to quake-hit areas

5.20

- Death toll rises to 40,075 as five million relocated
- Nation comes to 3-minute standstill
- Three counties have no power
- Most telecom services resume in quake-hit areas
- Crisis management ideas come of age
- Landslides hamper repair work
- Survivors of quake cope with thousands of aftershocks

5.21

- Toll passes 40,000, over 32,000 still missing
- Homes ready to house orphans, many keen on adoption
- China steps up battle to prevent epidemics

5.22

- They carry a lifeline through the air
- Restoring water supply "arduous task"
- Power, water back for most quake regions
- Injured get warm welcome, free treatment in other cities

5.23

- Quake toll rises to 55,239 in Sichuan
- China Mobile to restore service soon for quake relief

- Counseling service in quake areas
- Tens of thousands of Chinese seek to adopt orphans

5.24

- Wooden house proved lifesavers for thousand of Tibetans
- Quake-damaged railway reopens to traffic
- 10,000 quake-affected to be relocated to other provinces
- UN chief visits quake flattened town, offering support

5.25

- Premier says quake relief shifting to reconstruction
- 1.5 million houses to be built for quake-affected

5.26

- 182 aftershocks above 4 monitored in Sichuan
- Wolong pandas start life anew
- Strong aftershock destroys 71,000 homes, 6 killed

5.27

- Top legislator goes to epicenter by helicopter
- Volunteers embark on journey of self-discovery
- Magnitude 5.7 aftershock hits Shaanxi
- Ministry urges proper disposal of garbage in quake zone
- Pakistan medical team leaves for China's Gansu Province

5.28

- Official dismisses rumors of new massive quake
- China earthquake toll rises to 68,109

5.29

- Documentary to record earthquake stories
- Provincial authority outlines school collapse reasons
- Quake no big hurdle to economic growth

5.30

- New strategies needed to battle future shocks

5.31

- Mass evacuation underway for fear of "quake lake" burst

6.1

- Helicopter crashes while on duty in quake zones

6.2

- Telecommunications resume in hard-hit areas
- Third batch of Israeli relief supplies arrives in China
- Province-to-county support

6.3

- UAE to provide 50 million dollars in aid to quake-hit China
- Wenchuan's tragedy on film
- Children to be vaccinated against epidemics
- First school in epicenter resumes classes

6.4

- Sichuan maps tourism revival
- Quake inflation to be temporary
- Secondary disasters displace 30,000 people
- Regulation to direct post-quake reconstruction

6.5

- Insurers pay 233 million yuan in quake-related claims
- Serving people core of human rights practice
- Minister: food security guaranteed after quake
- Flood possible as "quake lake" nears drainage point

6.6

- China audits use of goods, funds in quake relief
- Overseas-funded companies donate 3.6 billion yuan

6.7

- College entrance exam starts as quake-zone students get grace period
- Kids back to school in quake-hit Wenxian

6.9

- China quake death toll rises to 69,142, missing down by 135
- China promulgates rules on post-quake reconstruction

6.10

- Experts choose new site for quake-leveled county
- Men on front more in need of counseling
- Qiang artists sell paintings to fund reconstruction

6.11

- Post-quake rebuilding rule focuses on public structures

6.12

- Bandengqiao likely site for new Beichuan

6.13

- Crematoriums worked round clock to cope

6.16

- Quake jeopardizes cultural heritage

6.17

- Quake relief HQ urges preventing secondary disasters

6.24

- Hu calls for enhanced disaster monitoring, early-warning capacity
- Main "quake lake" to be scenic spot
- Death toll in China's 12 May quake to exceed 80,000

6.25

- Major panda reserve resumes field inspection

6.28

- Little heroes awarded for quake rescue efforts

6.29

- Concert held in New York for sympathy for China's quake victims

7.1

- China quake death toll stands at 69,195

7.10

- World Bank approves 1 million dollars fund to help China identify contamination

7.11

- Foreign Direct Investment (FDI) grows 107.9% in Sichuan despite quake

7.14

- French medical team helps quake injured

7.16

- HK proposes 2b HKD for Sichuan reconstruction
- NPC Deputy proposes "National Disaster Prevention Day" on 12 May

7.18

- 2 Du Fu Thatched Cottage Museum receives donation
- Four sites nominated for preservation as quake museum

7.29

- Qiang culture fights for world heritage status

8.1

- Moderate tremor struck 12 May quake epicenter

8.6

- Death toll from Tuesday's aftershock rises to three

8.9

- "People-first" plan to govern Sichuan reconstruction

8.12

- 10 million quake survivors move into prefabs

8.17

- Post-quake rebuilding draft put online

8.22

- UN raises 15 million dollars to aid China quake zone
- School curriculums to add psychological counseling in quake-hit Sichuan

8.23

- China starts adoption process for 88 earthquake orphans

9.4

- Over 87,000 feared dead in 12 May earthquake

9.5
- Heritage restoration begins in quake-hit Sichuan

9.27
- Heavy rain lash quake zones in Sichuan, 16 killed

10.9
- Beijing to invest 7 billion yuan in quake-hit province

10.28
- 1.4 billion dollars plea to rebuild Qiang's treasures

11.12
- The first Wenchuan earthquake memorial park opened

11.14
- China receives 60 billion yuan donations for quake relief

11.21
- China publishes identities of 19,000 dead in May quake

12.1
- Tent-dwelling quake survivors move to temporary housing for winter

12.24
- 98 Russians awarded medals for Wenchuan quake rescue

Appendix 2

Wenchuan earthquake sequence catalog

From 12 May 2008 to 12 May 2009, M ≥ 4.0

Serial number	Time	Latitude	Longitude	Depth (km)	Magnitude
1	2008-05-12 14:43:15.1	31.0	103.5	33	6.0
2	2008-05-12 15:34:47.8	31.0	103.5	10	5.0
3	2008-05-12 15:40:05.7	31.0	103.6	10	4.7
4	2008-05-12 16:10:56.9	31.2	103.4	9	4.8
5	2008-05-12 16:21:47.3	31.3	104.1	33	5.2
6	2008-05-12 16:26:13.1	31.5	103.8	10	4.3
7	2008-05-12 16:35:03.4	31.4	103.5	10	4.6
8	2008-05-12 16:36:25.6	31.0	103.2	30	4.3
9	2008-05-12 16:47:28.9	32.2	105.4	33	4.8
10	2008-05-12 16:50:41.5	32.6	105.2	10	4.5
11	2008-05-12 17:01:48.4	32.2	104.7	10	4.1
12	2008-05-12 17:07:03.1	31.3	103.8	33	5.0
13	2008-05-12 17:23:38.5	32.3	104.8	10	5.1
14	2008-05-12 17:30:57.1	32.4	104.9	10	4.1
15	2008-05-12 17:42:26.4	31.4	104.0	33	5.2
16	2008-05-12 17:54:49.5	31.0	103.2	25	4.3
17	2008-05-12 18:02:29.6	32.2	105.1	10	4.8
18	2008-05-12 18:23:39.1	31.0	103.3	16	4.9
19	2008-05-12 19:10:58.4	31.4	103.6	33	6.0
20	2008-05-12 19:33:18.6	32.6	105.4		4.5
21	2008-05-12 19:41:10.3	32.4	105.1		4.6
22	2008-05-12 19:45:16.4	32.4	105.0		4.0
23	2008-05-12 19:52:22.3	32.6	105.4	15	4.7
24	2008-05-12 20:04:39.2	32.6	105.2	15	4.2
25	2008-05-12 20:06:33.1	32.2	105.5	15	4.1
26	2008-05-12 20:11:56.2	31.4	103.8	30	4.5

(continued)

Serial number	Time	Latitude	Longitude	Depth (km)	Magnitude
27	2008-05-12 20:15:41.5	32.0	104.4		4.9
28	2008-05-12 20:23:28.7	32.7	105.3	33	4.5
29	2008-05-12 20:29:56.5	31.4	103.9		4.1
30	2008-05-12 20:33:08.8	31.4	104.1	3	4.2
31	2008-05-12 20:54:56.7	31.3	103.4	23	4.3
32	2008-05-12 21:02:04.1	31.1	103.5	5	4.6
33	2008-05-12 21:07:15.2	31.0	103.4	5	4.3
34	2008-05-12 21:32:13.1	31.2	103.9	1	4.3
35	2008-05-12 21:36:05.0	32.9	105.5		4.0
36	2008-05-12 21:40:54.3	31.0	103.5	33	5.1
37	2008-05-12 21:55:14.0	32.0	104.3	3	4.2
38	2008-05-12 22:06:31.5	32.5	105.1		4.0
39	2008-05-12 22:09:46.8	31.9	104.7	1	4.5
40	2008-05-12 22:15:29.8	32.2	104.9	33	4.4
41	2008-05-12 22:26:15.8	31.3	103.9	14	4.0
42	2008-05-12 22:37:30.0	32.2	104.5	5	4.3
43	2008-05-12 22:46:09.5	32.7	105.5	33	5.1
44	2008-05-12 22:55:28.0	32.4	105.0	3	4.2
45	2008-05-12 23:05:29.3	31.3	103.6	17	5.0
46	2008-05-12 23:05:42.6	31.3	103.5	33	5.2
47	2008-05-12 23:16:57.7	30.9	103.2	20	4.1
48	2008-05-12 23:28:56.2	31.0	103.5	33	5.0
49	2008-05-13 00:28:54.6	31.2	103.8	33	4.4
50	2008-05-13 00:34:15.5	32.5	105.0	33	4.2
51	2008-05-13 01:01:45.7	30.9	103.4	17	4.0
52	2008-05-13 01:29:05.9	31.3	103.4	28	4.6
53	2008-05-13 01:54:30.8	31.3	103.4	19	5.0
54	2008-05-13 02:46:14.6	32.4	105.0	33	4.4
55	2008-05-13 02:55:24.7	31.9	105.1	33	4.4
56	2008-05-13 03:53:16.4	31.3	103.6	15	4.3
57	2008-05-13 04:08:50.1	31.4	104.0	33	5.7
58	2008-05-13 04:45:31.7	31.7	104.5	33	5.2
59	2008-05-13 04:51:29.1	32.4	105.2	33	4.7
60	2008-05-13 05:08:11.6	31.3	103.2	12	4.5
61	2008-05-13 05:51:26.7	32.5	105.3		4.6
62	2008-05-13 06:19:28.0	31.9	104.2	30	4.1
63	2008-05-13 06:24:36.2	32.2	105.0		4.0
64	2008-05-13 06:47:19.5	31.3	103.4	10	4.5
65	2008-05-13 07:38:13.6	31.9	104.5		4.0
66	2008-05-13 07:46:22.6	31.2	103.4	33	5.3
67	2008-05-13 07:54:46.4	31.3	103.6	33	5.1
68	2008-05-13 08:22:15.7	31.3	104.0	10	4.2

(continued)

Serial number	Time	Latitude	Longitude	Depth (km)	Magnitude
69	2008-05-13 08:54:04.9	32.6	105.2	33	4.3
70	2008-05-13 09:07:57.8	31.4	103.7	24	4.0
71	2008-05-13 10:15:15.7	31.6	103.9	30	4.5
72	2008-05-13 10:33:38.9	31.3	103.6	9	4.2
73	2008-05-13 10:59:31.1	31.0	103.3	22	4.3
74	2008-05-13 11:00:38.4	31.2	103.5	15	4.7
75	2008-05-13 11:48:46.5	31.2	103.7	26	4.5
76	2008-05-13 12:45:55.6	31.0	103.3	21	4.1
77	2008-05-13 12:50:21.8	31.3	103.4	21	4.1
78	2008-05-13 13:25:50.6	32.6	105.2	33	4.2
79	2008-05-13 13:36:34.5	32.4	105.2	33	4.4
80	2008-05-13 13:37:17.8	31.0	103.5	10	4.6
81	2008-05-13 14:38:17.6	31.4	103.8	21	4.3
82	2008-05-13 15:07:10.9	30.9	103.4	33	6.1
83	2008-05-13 15:19:16.4	32.3	105.0	33	4.8
84	2008-05-13 15:51:32.7	32.5	105.3	33	4.4
85	2008-05-13 15:53:03.6	32.3	105.0	33	4.8
86	2008-05-13 16:11:01.2	32.5	105.2	33	4.1
87	2008-05-13 16:20:49.3	31.4	103.9	0	4.8
88	2008-05-13 17:41:19.1	32.1	104.4	8	4.1
89	2008-05-13 18:16:06.1	31.8	104.3	0	4.0
90	2008-05-13 18:36:40.8	31.3	103.6	17	4.1
91	2008-05-13 20:51:36.4	32.3	105.0	0	4.6
92	2008-05-13 21:13:02.5	32.5	105.5	33	4.4
93	2008-05-13 21:31:32.3	32.4	105.1	33	4.5
94	2008-05-13 23:10:31.7	32.6	105.5	33	4.1
95	2008-05-13 23:54:27.5	32.1	104.9	1	4.5
96	2008-05-14 00:23:49.4	31.7	104.3	6	4.0
97	2008-05-14 03:30:16.1	31.1	103.3	13	4.1
98	2008-05-14 03:51:51.8	31.0	103.3	24	4.2
99	2008-05-14 06:03:29.8	31.3	103.6	10	4.2
100	2008-05-14 08:08:11.9	31.1	103.4	13	4.0
101	2008-05-14 09:09:19.4	31.4	103.8	10	4.8
102	2008-05-14 09:56:43.8	31.1	103.5	33	4.1
103	2008-05-14 10:54:36.5	31.3	103.4	33	5.6
104	2008-05-14 13:54:59.2	31.9	104.2	33	4.9
105	2008-05-14 14:33:55.2	31.4	103.9	12	4.1
106	2008-05-14 17:26:44.0	31.4	104.0	33	5.1
107	2008-05-14 17:51:18.2	32.4	104.2	95	4.8
108	2008-05-14 17:57:25.3	32.3	104.8		4.2
109	2008-05-14 18:00:28.4	32.2	104.7	10	4.7
110	2008-05-14 18:11:10.9	32.2	104.5	10	4.2

(continued)

Serial number	Time	Latitude	Longitude	Depth (km)	Magnitude
111	2008-05-14 18:18:15.1	32.4	105.1	33	4.8
112	2008-05-14 18:30:44.3	32.4	105.2	16	4.9
113	2008-05-14 21:29:45.1	32.3	105.1	33	4.5
114	2008-05-15 01:17:22.3	31.5	103.8	13	4.8
115	2008-05-15 01:33:24.0	31.4	103.5	33	4.6
116	2008-05-15 03:59:00.4	31.1	103.5	17	4.2
117	2008-05-15 05:01:08.0	31.6	104.2	33	5.0
118	2008-05-15 06:10:08.1	31.2	103.6	30	4.6
119	2008-05-15 08:09:28.5	31.8	104.4	21	4.3
120	2008-05-15 08:50:18.7	31.3	103.4	24	4.0
121	2008-05-15 12:27:29.1	31.3	103.7	22	4.1
122	2008-05-15 13:27:47.7	32.0	104.3	0	4.6
123	2008-05-15 20:10:23.9	31.4	103.8	18	4.2
124	2008-05-15 21:04:08.1	32.6	105.6	33	4.4
125	2008-05-16 05:55:47.4	32.3	104.7	0	4.5
126	2008-05-16 06:10:33.7	31.4	103.9	30	4.6
127	2008-05-16 06:34:32.1	31.9	104.4	17	4.2
128	2008-05-16 11:10:00.6	32.5	105.1	33	4.0
129	2008-05-16 11:34:28.5	31.4	104.1	33	4.9
130	2008-05-16 13:25:49.0	31.4	103.2	33	5.9
131	2008-05-16 14:34:39.6	32.4	105.2	33	4.3
132	2008-05-16 18:17:13.3	31.3	103.5	20	4.3
133	2008-05-16 18:20:49.1	32.5	105.1	33	4.0
134	2008-05-16 18:51:44.2	31.4	103.6	12	4.2
135	2008-05-16 21:21:10.3	31.9	104.2	30	4.0
136	2008-05-17 00:14:46.3	31.2	103.5	33	5.1
137	2008-05-17 01:22:20.4	31.2	103.6	9	4.5
138	2008-05-17 03:59:06.5	31.0	103.5	10	4.3
139	2008-05-17 04:00:13.9	32.6	105.4	33	4.1
140	2008-05-17 04:15:23.8	32.2	104.4	33	4.8
141	2008-05-17 04:16:52.0	31.3	103.5	33	5.0
142	2008-05-17 04:29:03.1	31.4	103.3	10	4.5
143	2008-05-17 06:33:11.3	32.2	105.1	33	4.2
144	2008-05-17 07:23:06.8	31.3	103.8	0	4.2
145	2008-05-17 08:28:59.0	31.6	104.0	30	4.1
146	2008-05-17 08:38:06.3	32.0	104.0	14	4.0
147	2008-05-17 15:38:43.2	32.0	104.4	10	4.1
148	2008-05-17 21:32:11.8	32.2	104.7	21	4.7
149	2008-05-18 01:08:23.4	32.1	105.0	33	6.0
150	2008-05-18 04:26:02.2	31.2	103.5	28	4.1
151	2008-05-18 08:45:55.7	31.8	104.0	10	4.2
152	2008-05-18 09:04:57.0	31.1	103.5	10	4.2

(continued)

Serial number	Time	Latitude	Longitude	Depth (km)	Magnitude
153	2008-05-18 11:51:42.7	31.0	103.4	6	4.2
154	2008-05-18 17:25:06.2	31.2	103.1	0	4.0
155	2008-05-18 20:37:01.5	31.3	103.2	21	4.2
156	2008-05-19 12:08:58.9	32.1	105.0	33	4.8
157	2008-05-19 14:06:54.9	32.5	105.3	33	5.4
158	2008-05-20 01:52:36.4	32.3	104.9	33	5.0
159	2008-05-20 08:57:36.1	31.7	104.0	5	4.1
160	2008-05-20 11:42:28.6	32.6	105.4	33	4.0
161	2008-05-20 12:17:56.3	30.8	103.3	18	4.3
162	2008-05-20 14:54:21.2	31.8	104.2	5	4.0
163	2008-05-21 00:38:37.9	30.9	103.3	21	4.0
164	2008-05-21 16:40:51.4	31.4	103.3	17	4.1
165	2008-05-21 17:33:13.6	32.3	105.2	33	4.5
166	2008-05-21 21:59:06.9	31.4	103.9	30	4.5
167	2008-05-21 23:29:52.6	32.4	105.1	33	4.3
168	2008-05-22 04:36:26.9	32.2	104.8	20	4.6
169	2008-05-22 15:18:42.7	31.2	103.6	30	4.7
170	2008-05-22 19:22:27.5	32.6	105.4	33	4.0
171	2008-05-22 23:00:11.7	31.9	104.3	7	4.3
172	2008-05-23 01:37:39.7	31.3	103.6	33	4.6
173	2008-05-23 08:05:04.1	31.2	103.6	33	4.7
174	2008-05-23 09:23:27.7	31.2	103.5	20	4.2
175	2008-05-23 11:12:13.5	31.2	103.2	20	4.2
176	2008-05-24 00:10:24.7	32.2	105.0	33	4.0
177	2008-05-24 01:53:34.5	32.5	105.2	33	4.1
178	2008-05-24 11:00:12.1	31.1	103.4	7	4.0
179	2008-05-25 12:27:05.2	32.0	104.6	12	4.2
180	2008-05-25 16:21:46.9	32.6	105.4	33	6.4
181	2008-05-25 17:34:07.1	33.0	104.9	33	4.7
182	2008-05-26 08:39:47.4	30.8	103.3	21	4.3
183	2008-05-27 16:03:24.1	32.7	105.6	33	5.4
184	2008-05-27 16:37:53.1	32.8	105.6	33	5.7
185	2008-05-27 21:59:36.1	32.5	105.2	33	4.7
186	2008-05-28 00:46:06.7	32.2	104.6	20	4.2
187	2008-05-28 01:35:11.2	32.7	105.4	33	4.7
188	2008-05-29 12:48:43.3	32.6	105.5	33	4.6
189	2008-05-29 15:10:20.1	31.4	103.7	20	4.5
190	2008-05-31 14:22:42.8	32.4	105.0	33	4.0
191	2008-05-31 15:34:21.9	32.6	105.4	33	4.0
192	2008-06-01 11:23:57.5	31.6	104.0	21	4.5
193	2008-06-03 11:09:28.4	32.0	104.5	6	4.3
194	2008-06-05 01:26:41.3	32.3	105.1	33	4.0

(continued)

Serial number	Time	Latitude	Longitude	Depth (km)	Magnitude
195	2008-06-05 05:21:39.1	31.2	103.4	12	4.2
196	2008-06-05 12:41:08.2	32.3	105.0	33	5.0
197	2008-06-05 14:02:32.2	32.7	105.5	33	4.3
198	2008-06-06 19:03:48.7	31.3	103.8	23	4.0
199	2008-06-06 22:38:46.6	31.1	103.3	29	4.0
200	2008-06-07 08:48:12.1	31.2	103.5	33	4.6
201	2008-06-07 14:28:34.8	32.5	105.4	33	4.3
202	2008-06-07 15:32:47.8	32.5	105.2	33	4.0
203	2008-06-07 18:42:04.7	32.4	105.6	33	4.6
204	2008-06-08 06:14:29.0	32.5	105.1	33	4.7
205	2008-06-08 18:51:19.5	31.9	104.4	33	4.8
206	2008-06-09 15:28:36.0	31.4	103.8	33	5.0
207	2008-06-11 00:27:28.6	32.8	105.8	33	4.3
208	2008-06-11 06:23:20.2	30.9	103.4	33	5.0
209	2008-06-15 08:11:20.7	31.3	103.6	10	4.5
210	2008-06-16 12:31:39.3	31.3	103.7	17	4.1
211	2008-06-17 13:14:33.7	31.7	104.1	18	4.1
212	2008-06-17 13:51:44.8	32.8	105.6	33	4.5
213	2008-06-17 21:32:06.2	31.9	104.3	17	4.0
214	2008-06-17 21:40:46.8	32.3	104.9	33	4.0
215	2008-06-18 06:43:13.4	31.3	103.4	14	4.0
216	2008-06-19 10:55:29.1	31.2	103.5	17	4.3
217	2008-06-19 12:48:16.6	31.8	104.3	16	4.2
218	2008-06-19 18:25:58.2	32.8	105.5	33	4.4
219	2008-06-20 04:27:46.5	31.2	103.5	9	4.7
220	2008-06-22 18:37:34.0	32.2	104.5	18	4.2
221	2008-06-23 05:38:32.4	32.4	105.1	10	4.1
222	2008-06-27 08:55:40.8	31.1	103.4	16	4.1
223	2008-06-28 02:20:53.6	31.5	103.2	14	4.6
224	2008-06-28 05:42:11.3	32.3	104.9	21	4.6
225	2008-06-29 07:55:20.0	32.1	104.6	22	4.1
226	2008-07-02 03:57:53.5	31.4	103.9	17	4.1
227	2008-07-03 12:08:12.0	31.2	103.5	16	4.0
228	2008-07-05 06:10:54.6	31.8	104.1	14	4.1
229	2008-07-05 15:00:44.5	31.6	104.1	11	4.5
230	2008-07-06 02:50:13.6	31.6	104.1	12	4.0
231	2008-07-06 17:37:50.3	31.2	103.6	15	4.0
232	2008-07-08 20:12:58.1	31.4	103.9	8	4.3
233	2008-07-10 07:19:50.0	32.2	104.9	20	4.0
234	2008-07-15 03:56:33.2	31.6	104.0	29	4.2
235	2008-07-15 17:26:21.5	31.6	104.0	15	5.0
236	2008-07-18 00:40:41.3	31.7	104.1	10	4.6

(continued)

Serial number	Time	Latitude	Longitude	Depth (km)	Magnitude
237	2008-07-18 09:26:01.9	32.4	105.2	10	4.1
238	2008-07-24 03:54:46.5	32.8	105.6	10	5.6
239	2008-07-24 13:30:08.8	32.8	105.5	10	4.9
240	2008-07-24 15:09:28.6	32.8	105.5	10	6.0
241	2008-07-25 04:54:15.7	30.8	103.3	15	4.4
242	2008-07-25 21:39:51.5	32.8	105.5	10	4.2
243	2008-07-29 07:52:07.7	31.3	103.8	17	4.1
244	2008-08-01 16:32:44.6	32.1	104.7	20	6.1
245	2008-08-02 02:12:17.5	32.5	105.2	10	4.9
246	2008-08-02 21:25:49.5	32.7	105.6	10	4.2
247	2008-08-02 21:35:59.6	32.6	105.4	10	4.1
248	2008-08-05 17:49:18.7	32.8	105.5	10	6.1
249	2008-08-06 07:55:06.6	32.8	105.5	15	4.5
250	2008-08-06 11:42:26.2	32.7	105.4	10	4.4
251	2008-08-06 12:47:05.3	32.8	105.5	10	4.3
252	2008-08-07 16:15:33.7	32.1	104.7	10	5.0
253	2008-08-08 21:12:19.3	31.5	103.9	21	4.3
254	2008-08-09 20:10:19.6	32.7	105.4	10	4.3
255	2008-08-10 00:12:35.6	32.4	105.1	16	4.0
256	2008-08-10 00:31:36.0	32.5	105.2	10	4.2
257	2008-08-10 12:38:09.4	32.4	105.1	10	4.0
258	2008-08-13 05:03:21.2	31.9	104.2	18	4.9
259	2008-08-13 16:45:42.4	32.5	105.4	10	4.0
260	2008-08-15 01:06:38.5	31.0	103.2	20	4.9
261	2008-08-18 10:22:22.8	31.3	103.4	15	4.0
262	2008-09-11 10:53:45.2	32.4	105.2	15	4.4
263	2008-09-11 18:30:21.8	31.3	103.7	17	4.3
264	2008-09-12 01:38:58.6	32.9	105.6	6	5.5
265	2008-09-20 00:29:28.8	31.3	103.5	14	4.1
266	2008-10-04 20:11:46.6	31.8	104.1	21	4.5
267	2008-10-04 23:55:33.3	31.5	103.9	21	4.0
268	2008-10-05 02:28:49.5	30.9	103.2	24	4.0
269	2008-10-16 10:08:44.7	31.3	103.5	15	4.1
270	2008-10-16 17:46:16.7	31.2	103.7	15	4.1
271	2008-10-21 15:25:28.7	31.7	104.1	16	4.1
272	2008-10-24 05:56:04.4	32.5	105.2	14	4.0
273	2008-10-25 01:50:33.8	31.4	103.9	21	4.1
274	2008-10-31 10:06:11.2	31.8	104.4	10	4.3
275	2008-11-04 20:43:13.5	32.1	104.5	22	4.5
276	2008-11-06 23:01:41.6	31.0	103.5	16	4.0
277	2008-11-14 11:27:05.4	32.0	103.0	11	4.6
278	2008-11-14 14:33:34.4	32.8	105.4	7	4.3

(continued)

Serial number	Time	Latitude	Longitude	Depth (km)	Magnitude
279	2008-11-16 06:59:49.5	32.2	104.7	22	5.1
280	2008-11-19 07:07:24.8	31.7	104.2	16	4.1
281	2008-11-23 18:19:01.8	31.2	103.5	16	4.7
282	2008-12-04 21:57:39.6	31.4	103.4	15	4.1
283	2008-12-07 04:26:02.7	31.2	103.5	17	4.0
284	2008-12-07 11:02:20.6	31.7	104.4	1	4.2
285	2008-12-10 02:53:11.5	32.6	105.4	15	5.0
286	2008-12-23 08:33:17.8	31.3	103.2	10	4.2
287	2008-12-29 14:18:21.8	32.3	105.0	19	4.9
288	2009-01-02 19:00:10.4	31.9	104.2	19	4.7
289	2009-01-07 13:22:17.1	32.1	104.4	20	4.2
290	2009-01-15 02:23:36.2	31.3	103.3	22	5.1
291	2009-02-01 01:36:38.7	32.5	105.3	17	4.5
292	2009-02-08 05:21:05.2	30.9	103.2	22	4.3
293	2009-02-08 23:17:10.4	31.4	103.8	18	4.2
294	2009-03-04 00:21:46.4	31.9	104.8	10	4.2
295	2009-03-12 16:25:38.0	32.4	105.0	10	4.7
296	2009-03-19 13:43:30.6	32.3	105.0	20	4.2
297	2009-04-06 00:47:03.8	31.9	104.2	17	4.6
298	2009-05-14 23:49:28.7	32.3	104.8	13	4.2
299	2009-06-30 02:03:50.5	31.5	104.0	20	5.6
300	2009-06-30 13:40:24.7	31.5	104.0	18	4.3
301	2009-06-30 15:22:19.7	31.5	104.0	20	5.0
302	2009-07-17 22:35:41.1	31.4	103.9	20	4.5
303	2009-08-01 05:54:38.2	31.8	104.1	18	4.1
304	2009-08-04 06:18:13.2	31.2	103.6	16	4.0
305	2009-09-11 22:36:29.6	31.5	103.8	18	4.0
306	2009-09-19 16:54:14.3	32.8	105.6	7	5.1
307	2009-10-20 00:42:43.5	32.1	104.5	23	4.9
308	2009-10-29 21:28:00.5	32.6	105.2	10	4.8
309	2009-11-23 15:25:21.7	31.0	103.2	21	4.8
310	2009-11-28 00:04:02.6	31.3	103.9	21	5.0
311	2009-12-05 01:43:40.9	32.1	104.2	16	4.0
312	2010-02-27 05:37:32.2	31.2	103.4	12	4.2
313	2010-03-23 23:52:21.3	31.3	103.4	17	4.0

Appendix 3

Law of the People's Republic of China on Protecting Against and Mitigating Earthquake Disasters

Order of the President of the People's Republic of China, No. 7

The Law of the People's Republic of China on Protecting Against and Mitigating Earthquake Disasters, amended and adopted at the 6th Meeting of the Standing Committee of the Eleventh National People's Congress of the People's Republic of China on December 27, 2008, is hereby promulgated and shall go into effect as of May 1, 2009.

Hu Jintao
President of the People's Republic of China
December 27, 2008

Contents

Chapter I	General Provisions
Chapter II	Plans for Protecting Against and Mitigating Earthquake Disasters
Chapter III	Earthquake Monitoring and Prediction
Chapter IV	Protecting Against Earthquake Disasters
Chapter V	Earthquake Emergency Rescue and Relief
Chapter VI	Post-Earthquake Transitional Resettlement, Rehabilitation and Reconstruction
Chapter VII	Supervision and Administration
Chapter VIII	Legal Responsibility
Chapter IX	Supplementary Provisions

(source: http://www.lawinfochina.com)

Appendix 4

The Overall Planning for Post-Wenchuan Earthquake Restoration and Reconstruction (Extracts)

State Council, 19 September 2008

Foreword

The present planning is hereby formulated for the purpose of doing a good restoration and reconstruction job in a powerful, orderly and efficient way and rebuilding a beautiful homeland so as to restore the normal socioeconomic order in the quake-hit areas.

Contents

Foreword
Chapter I   Basis for Reconstruction
Chapter II   General Requirements
Chapter III   Spatial Distribution
Chapter IV   Urban and Rural Housing
Chapter V   Urban Construction
Chapter VI   Rural Construction
Chapter VII   Public Services
Chapter VIII   Infrastructure
Chapter IX   Industrial Reconstruction
Chapter X   Disaster Prevention and Mitigation
Chapter XI   Eco-environment
Chapter XII   Spiritual Homeland
Chapter XIII   Policies and Measures
Chapter XIV   Reconstruction Funds
Chapter XV   Planning Implementation

(source: http://www.recoveryplatform.org)

Chapter I Basis for Reconstruction

The Wenchuan earthquake affected 417 counties (cities, districts) of 10 provinces (autonomous regions and municipalities) such as Sichuan, Gansu, Shaanxi, Chongqing, Yunnan Provinces, etc., covering approximately a total area of 500,000 km². The planned scope of the present planning includes 51 counties (cities and districts) in the extremely hard-hit and hard-hit disaster areas of Sichuan*, Gansu and Shaanxi Provinces, covering a total area of 132,596 km², involving 14,565 administrative villages in 1271 towns and townships, with a total population of 19.867 million by the end of 2007, the gross regional product of 241.8 billion yuan, urban household per capita disposable income and rural household per capita net income of 13,050 yuan, 3533 yuan, respectively.

Column 1 Planned Scope

Province	County (city and district)	No.
Sichuan	Wenchuan, Beichuan, Mianzhu City, Shifang City, Qingchuan, Maoxian, Anxian, Dujiangyan City, Pingwu, Pengzhou City, Lixian, Jiangyou, Lizhou (District of Guangyuan City), Chaotian (District of Guangyuan City), Wangcang, Zitong, Youxian (District of Mianyang City), Jingyang (District of Deyang City), Xiaojin, Fucheng (District of Mianyang City), Luojiang, Heishui, Chongzhou City, Jiange, Santai, Langzhong City, Yanting, Songpan, Cangxi, Lushan, Zhongjiang, Yuanba (District of Guangyuan City), Dayi, Baoxing, Nanjiang, Guanghan City, Hanyuan, Shimian, Jiuzhaigou	39
Gansu	Wenxian, Wudu (District of Longnan City), Kangxian, Chengxian, Huixian, Xihe, Liangdang, Zhouqu	8
Shaanxi	Ningqiang, Lueyang, Mianxian, Chencang (District of Baoji City)	4

Chapter II General Requirements

Achieve the major task of restoration and reconstruction in approximately three years. The basic living conditions and the economic development level should reach or surpass the pre-disaster level. Make every endeavor to build a new homeland characterized by enjoyable life and work, eco-civilization, security and harmony, and lay a solid foundation for sustainable socioeconomic development.

---

* The scope of the extremely hard-hit and hard-hit areas is determined in accordance with *Evaluation Results of the Wenchuan Earthquake Disaster Areas by Ministry of Civil Affairs*, etc.

- Housing shall be available to each family. Basically complete the restoration and reconstruction of urban and rural residences, make it possible for the disaster-affected people to live in safe, economical, practical and land-saving houses.
- Ensure employment for household of working population. Ensure that at least one member in each family has a stable job, and that urban household per capita disposable income and rural household per capita net income surpass the pre-disaster level.
- Each person is secured. All the disaster-affected people shall enjoy basic living allowances, and basic public services such as compulsory education, public sanitation and basic medical treatment, and have access to public culture and sports, social welfare, etc.
- Improve infrastructures. Completely restore the functions of infrastructures such as transportation, communications, energy, water conservancy, etc., ensure that the support capacity reach or surpass the pre-disaster level.
- Ensure development in economy. Improve and expand special advantage industries, optimize industry structure and spatial layout and enhance the scientific development capacity.
- Ensure improvement in ecology. Gradually restore the ecological functions and improve environmental quality and ensure apparent enhancement in disaster prevention and mitigation ability.

Chapter III Spatial Distribution

On the basis of the comprehensive evaluation of the resources and environment carrying capacity and in accordance with the national land development intensity, industry development direction as well as the appropriateness of population aggregation and urban construction, the national land in the planned areas is divided into three categories: suitable for reconstruction; appropriate reconstruction; and ecological restoration[**].

Column 2 Reconstruction Divisions[***]

Type	Area (km²)	Proportion in the planned areas (%)	Population (×10⁴ people)	Proportion in the planned areas (%)
Areas suitable for reconstruction	10,077	7.6	772.8	38.9
Areas suitable for appropriate reconstruction	38,320	28.9	1180.1	59.4
Ecological reconstruction area	84,199	63.5	33.8	1.7

1. Areas Suitable for Reconstruction
   - Mainly refers to the areas with relatively strong resources and environment carrying capacity and smaller disaster risks, suitable for the reconstruction of county seats, towns and townships on the original sites, for the aggregation of a relatively large population, and for the overall development of various industries.
   - Mainly distributed in the piedmont plains of the Longmen Mountains and the shallow hill areas bordering the said mountains in Sichuan Province, in the river valleys of Weihe River and Jinghe River and Huicheng Basin in Gansu Province, on the edge of Hanzhong Basin and the transitional belt of Guanzhong Plain in Shaanxi Province, as well as a few other scattered plots.
   - The functions are oriented to promote industrialization and urbanization, to aggregate population and economy and to build into the zones for revitalizing economy, carrying industries and creating employment. The areas suitable for reconstruction in Sichuan, Gansu and Shaanxi Provinces shall become important components of Cheng (du)-De (yang)-Mian(yang) Economic Zone, Tianshui Economic Zone and Guanzhong Economic Zone, respectively.
2. Areas Suitable for Appropriate Reconstruction
   - Mainly refer to the areas with relatively weak resources and environment carrying capacity and comparatively big disaster risks, suitable for appropriate reconstruction of county seats, towns and townships on the original sites under the precondition of controlled scale, for appropriate population aggregation, and for the development of specific industries.
   - Mainly distributed in the altiplano area behind Longmen Mountains and canyons within mountains in Sichuan Province, the west Qinling mountainous area in Gansu Province, the Qinba mountainous area in Shaanxi Province and other areas where the development intensity should be controlled.
   - The functions are oriented to give priority to protection, and carry out

---

** The scope and size of the reconstruction divisions are ascertained on the basis of the *Evaluation Report on Resources and Environment Carrying Capacity* issued by Chinese Academy of Sciences.

*** The data and areas of the division of reconstruction areas in the column are the results of real calculation according to the natural distribution. The population is obtained after data processing on the basis of township statistics.

appropriate exploitation and spotty development, so as to build the zones with appropriate population, good eco-environment and distinctive industrial characteristics.

3. Ecological Reconstruction Areas
   - Mainly refers to the areas with very low resources and environment carrying capacity, great disaster risks and significant ecological functions, where the construction land is in severe shortage and the cost of construction and maintenance of transportation and other infrastructures is extremely high, and where it is inappropriate to reconstruct towns in the original sites or to aggregate a large population.
   - Mainly distributed in the core area of the Longmen Mountains quake fault zone and alpine areas in Sichuan Province, the Kuma-Longmen Mountains fault zone in Gansu Province, Mianlue fault zone in Shaanxi Province and the various conservation zones at all levels and of all categories.
   - The functions are oriented to focus on the ecological protection and restoration, and build the areas for protecting natural and cultural resources as well as rare and precious fauna and flora resources, with a small scattered population.

Chapter IV Urban and Rural Housing

Corresponding governmental subsidiary policies shall be formulated on the basis of different characteristics of urban and rural housing construction and consumption in the process of urban and rural housing restoration and reconstruction. Those reparable houses which could ensure security are not necessary to be destroyed and rebuilt. Immediate check and inspection, timely maintenance and strengthening are necessary in such cases. Scientific site option, intensive and economical land use and seismic fortification criteria confirmation and execution are essential for the implementation of construction of those houses in need of reconstruction.

Chapter V Urban Construction

The restoration and reconstruction of towns shall follow the requirements of restoring and perfecting functions, comprehensive considerations and arrangements to optimize the layout of urban space, enhance the disaster prevention ability, and improve residential environment, laying a foundation for the sustainable development for the towns.

Column 6 Municipal Public Facilities

Field	Projects	Total Restoration	Total New construction	Sichuan Restoration	Sichuan New construction	Gansu Restoration	Gansu New construction	Shaanxi Restoration	Shaanxi New construction
Transportation	Road (km)	2548	1509	2301	1332	180	94	67	83
	Bridge	728	123	635	58	54	22	39	43
	Bus station	450	207	419	130	24	3	7	74
Water supply	Water plant	451	15	442	12	8	—	1	3
	Pipe net (km)	4153	2363	4055	2085	74	119	24	159
Gas and heat supply	Gas distribution station	203	15	203	10	2	—	—	3
	Gas supply network (km)	2052	791	2049	590	—	—	3	201
	Heating-electricity plant	3	4	—	—	3	4	—	—
	Gas supply network (km)	6	41	—	—	6	41	—	—
Sewage disposal	Treatment plant	331	27	328	21	3	3	—	3
	Pipe net (km)	800	7256	765	6350	29	471	6	435
Garbage disposal	Treatment plant	47	8	39	1	5	5	3	2
	Transfer station	665	87	565	9	44	60	56	18

Column 8 Production Facilities and Bases of Agriculture

Production facilities and bases of agriculture
Restore 100,500 hectares of damaged farmland, restore and reconstruct 28.8 million m^2 of greenhouse, 22.11 million m^2 of livestock and poultry houses, 12,300 hectares of aquaculture pond, and 9982 electromechanical lift irrigation stations, 18,392 km farm machinery road.

Production bases of premium grain and oil
Construct 20 rice production bases, 14 corn production bases, 21 potato production bases, 23 "dual low" rape production bases, and 7300 hectares of olive growing bases.

Production bases of high valued fruit and vegetable
Construct 33 vegetable production bases, 18 production bases of high valued fruit and vegetable and 13 production bases of edible fungi.

Production bases of tea, drug and mulberry
Construct 13 production bases of tea, 23 production bases of Chinese traditional medicine, and 28 production bases of sericulture industry.

Production bases of stock raising
Construct production base that can produce 8.9 million hogs every year, production base that can produce 2.26 million mutton sheep every year, production base that can produce 420 thousand beef cattle every year, milk production base that can hold 42 thousand livestock, production base that can produce 8 million native chicken, production base that can hold 6.5 million rabbits, and production base that can produce 5000 tons of bee products.

Production base of aquatic products
Construct 39 characteristic aquiculture bases.

Production bases of forestry industry
Construct 19,300 hectares of raw material forests of wood and bamboos, and 15,300 hectares of walnut and other characteristic economic forest.

Chapter VI Rural Construction

The restoration and reconstruction of rural production and living facilities shall conform to the requirements of balancing urban and rural development, while combining new countryside construction and poverty alleviation and development so as to reach the purposes of resources integration, zoning design, hierarchy configuration, facilitating and benefiting residents and co-construction and sharing.

Chapter VII Public Services

Ensure resources integration and configuration readjustment in light of urban and rural layout and population size for public service facilities restoration and reconstruction, so as to promote standardized construction as well as the equalization of public service. Give priority to the restoration and reconstruction of schools, hospitals and other public service facilities while strictly abide by compulsory construction standard norms for the safest, soundest and most reliable architectures.

Chapter VIII Infrastructure

Priority shall be given to the function of restoration in the restoration and reconstruction of infrastructure. Make rational adjustment to the infrastructure distribution in accordance with geological and geographical conditions and rural and urban layout, integrated with local socioeconomic development planning, rural and urban planning and land utilization planning. Combine the resources, far and near, optimize the structure, rationally determine the construction criteria and enhance the safety assurance capacity.

Chapter IX Industrial Reconstruction

According to the resource environment carrying capacity, industrial policies and employment demands, the industrial restoration and reconstruction shall be market-oriented with enterprises as the main body. We shall appropriately guide the disaster-affected enterprises to conduct onsite restoration and reconstruction, reconstruction at separate locations, suspend operation and merge with other enterprises, support the special and advantage industries, so as to promote the restructuring, facilitate the change of development modes and expand employment.

Chapter X Disaster Prevention and Mitigation

The restoration and reconstruction of the disaster prevention and reduction system shall be in line with the principle of putting prevention first, reasonable avoidance, comprehensive treatment and overall management so as to enhance the construction of disaster prevention and mitigation system and the comprehensive disaster alleviation capacity, and to improve the disaster prevention and emergency assistance ability.

Chapter XI Eco-environment

The restoration and reconstruction of the ecological environment shall be in line with the principle of respecting nature, regulations and laws as well as science. In addition, ecological restoration and environmental governance shall be intensified with a view to promoting coordinated development of population, resources and environment.

Chapter XII Spiritual Homeland

The restoration and reconstruction of the spiritual homeland shall focus on following aspects: psychological guidance, enhancement of the humane care, remolding of the positive and optimistic mental outlook, strengthening the confidence in self-reliance and arduous struggle, as well as popularization of the great spirit of earthquake fighting and disaster relief and the excellent traditional culture of the Chinese nation.

Chapter XIII Policies and Measures

Adhere to the principle of special cases with special methods, formulate and implement policies and measures with strong pertinence in accordance with the needs of restoration and reconstruction, strengthen coordination and form composition forces so as to provide policy support for realizing the target determined in this planning and achieving the reconstruction task.
- Establish the fund for restoration and reconstruction. The central finance shall establish the fund for post-quake restoration and reconstruction. The fund for disaster-stricken areas mainly refers to the provincial finance established in contrast with the post-disaster restoration and

reconstruction fund.

- Alleviate the tax burden on individuals. The disaster-assistance and donations of every individual in disaster areas and the subsidies of the front-line staffs working for the earthquake fighting and disaster relief and for the restoration and reconstruction obtained in accordance with the regulated standards shall be exempted from the income tax.
- Make clear the assistance task. According to their annual material workload, 19 assistance provinces (cities) shall offer assistance with no less than 1% of their last ordinary budget revenues to their 24 counterpart counties (cities, districts) in Sichuan, Gansu and Shaanxi Provinces[IV].

Chapter XIV Reconstruction Funds

Insist on employing the reformed methods to raise the restoration and reconstruction funds through multiple channels, bring into full play the enthusiasm of all the parties concerned. Actively innovate the financing and utilizing approaches, enhance the capital utilization efficiency and improve funding management and supervision mechanism so as to provide the fund guarantee for achieving the targets prescribed in the present planning and fulfill the reconstruction task.

- According to the designated target and reconstruction task in this layout, the total capital demand for restoration and reconstruction is calculated approximately as 1000 billion RMB yuan.
- In light of the proportion of approximately 30% of aggregate demand of the restoration and reconstruction funds, the central finance shall establish the funds for post-quake restoration and reconstruction.
- Counterpart assistance funds shall be mainly applied to the restoration and reconstruction of urban and rural residents' housing, public services, municipal public facilities, agricultural and rural infrastructures as well as services such as planning formulation, architectural design, expert consultation, engineering construction and supervision, etc.

Chapter XV Planning Implementation

We should establish and strengthen the mechanism of Planning

IV Besides the 18 counties (cities) in Sichuan Province that have been promulgated to accept the counterpart assistance, Wenxian County, Wudu District, Kangxian County, Zhouqu County in Gansu Province and the counties of Niqiang and Lueyang in Shaanxi Province were added.

Implementation, definite aims and tasks, seize a proper time sequence of reconstruction, carry out the responsibilities, complete the supervision and assessment and actively promote the smooth implementation of the planning. The task is arduous, time is pressing and influence is profound for the post-Wenchuan quake restoration and reconstruction. Under the firm leadership of CPC Central Committee with Comrade Hu Jintao as general secretary, with the substantial support of our people of all nationalities, the Cadres and masses in disaster areas shall certainly rebuild a new homeland characterized by enjoyable life and work, eco-civilization, security and harmony with their industrious efforts.